QUANTITATIVE ASSESSMENT OF GROU
SURFACE WATER INTERACTIONS IN THE HAILIUTU RIVER
BASIN, ERDOS PLATEAU, CHINA

Zhi YANG

QUANTITATIVE ASSESSMENT OF GROUNDWATER AND SURFACE WATER INTERACTIONS IN THE HAILIUTU RIVER BASIN, ERDOS PLATEAU, CHINA

DISSERTATION

Submitted in fulfillment of the requirements of
the Board for Doctorates of Delft University of Technology
and
of the Academic Board of the IHE Delft
Institute for Water Education
for
the Degree of DOCTOR
to be defended in public on
Monday, 26 March 2018 at 10.00 hours
in Delft, the Netherlands

by

Zhi YANG
Master of Science in Hydrology and Water Resources,
UNESCO-IHE Institute of Water Education,
Delft, the Netherlands
born in Anhui Province, China

This dissertation has been approved by the
promotor: Prof. dr. S. Uhlenbrook
copromoter: Dr. Y. Zhou

Composition of the doctoral committee:

Rector Magnificus	Chairman
Prof. dr. ir. E.J. Moors	Vice-Chair IHE Delft
Prof. dr. S. Uhlenbrook	Delft University of Technology/IHE Delft
Dr. Y. Zhou	IHE Delft

Independent members:
Prof. dr. Z. Su	Twente University/ITC
Prof. dr. X. Wang	China University of Geosciences, Beijing, China
Prof. dr. ir. F.C. van Geer	Utrecht University
Prof. dr. ir. T.N. Olsthoorn	Delft University of Technology
Prof. dr. ir. M.E. McClain	IHE Delft/Delft University of Technology, reserve member

This research was conducted under the auspices of the Graduate School for Socio-Economic and Natural Sciences of the Environment (SENSE)

CRC Press/Balkema is an imprint of the Taylor & Francis Group, an informa business

Published by:
CRC Press/Balkema
Schipholweg 107C, 2316 XC, Leiden, the Netherlands
Pub.NL@taylorandfrancis.com
www.crcpress.com – www.taylorandfrancis.com
ISBN 978-1-138-59687-0

Summary

Groundwater and surface water have been considered as isolated components of the hydrological cycle for centuries in the application of water resources management. However, groundwater has hydraulic connections with the surface water system and forms a single resource. The interactions between groundwater and surface water comprise complex processes at many different temporal and spatial scales. A growing number of studies on the interactions between groundwater and surface water have been conducted from mountain to flood plain areas in tropical and temperate zones with arid and humid climate. Human activities have interfered with the natural connections between groundwater and surface water according to previous studies. The dynamics of interactions between groundwater and surface water, as well as the response to human interferences, draws our interest towards the influential effects.

The Hailiutu River catchment on the semi-arid Erdos plateau, China, is suffering from the conflicts between the exploitation of water resources for social and economic development and ecosystem protection. The local community has experienced water shortage, groundwater level decline, and ecosystem deterioration in the last several decades. Although groundwater plays a highly important role in the water supply for the local community and sustains the groundwater-dependent ecosystem in the Hailiutu catchment, the interactions between groundwater and surface water, as well as the consequences of human impacts on the hydrological processes, have not been thoroughly elucidated to date. The understanding of the mechanisms associated with the interactions between groundwater and surface water is crucial for achieving sustainable development.

This thesis presents a systematic approach for investigating the interactions between groundwater and surface water, which consists of a statistical analysis of water resources dynamics at the basin level, multiple in-situ observation methods for determining groundwater contributions at a local scale, a simple to complex water balance analysis at the sub-catchment scale, and the quantification of temporal and spatial interactions between groundwater and surface water for the Hailiutu River catchment with tracer and modelling methods.

According to the systemic study of the quantitative assessment of groundwater and surface water interactions in the Hailiutu River catchment, the primary contents and conclusions of the thesis are summarized as follows.

First, the importance of interactions between groundwater and surface water, as well as the human impacts at the semi-arid Erdos plateau, China, for sustainable development and ecosystem protection are emphasized. The general background information of the Erdos plateau and the selected Hailiutu River catchment are introduced. The research objectives are identified.

The shifts of stream flow regime at the Hailiutu River were systematically analysed to identify the impacts of different driving forces including climate and human activities. After statistically detecting changes in river discharges, five periods with four major shifts in the flow regime were observed in 1968, 1986, 1992 and 2001. The flow regime reflects the quasi-natural conditions of high variability and larger amplitude of 6-month periodic fluctuations in the period from 1957 to 1967. The peak flow of the river was reduced by the construction of two reservoirs in the period 1968 to1985. In the period 1986 to 1991, the river discharge further decreased due to the combined influence of river diversions and the increase of groundwater extractions for irrigation. In the fourth period from 1992 to 2000, the river discharge reached its lowest flow and variability, which corresponds to a large increase in crop area. The flow regime recovered in the fifth period from 2001 to 2007 with the implementation of the government policy of returning farmland to forest and grassland, which converts crop land into natural xeric bushland. We identified (or: It was concluded) that climatic elements such as precipitation and air temperature had minor impacts on the flow regime shifts. The construction of the hydraulic works for surface water diversion, groundwater extraction, and the land use policy changes were responsible for flow regime changes in the Hailiutu River catchment.

The second topic focuses on in-situ observations at local scale. The identification and quantification of the interactions between groundwater and surface water at Bulang tributary of the Hailiutu River were conducted by means of hydraulic, temperature, hydrochemistry and isotopic methods with field observations for one year (November 2010 to October 2011). The groundwater discharges to the river dominate the hydrological processes according to the field measurements. The observed groundwater levels were always higher than the river stage, which indicated the groundwater discharge to the river. Temperature measurements of stream water, streambed deposits at different depths, and groundwater confirm the upward flow of groundwater to the stream during all seasons. The results of a tracer-based hydrograph separation exercise reveal that, even during heavy rainfall events, groundwater contributes much more to the increased stream discharge than direct surface runoff. Furthermore, groundwater seepage along the reach was quantified with combined river discharge measurements and EC profile measurements under natural condition and constant injection.

To determine the temporal and spatial interactions between groundwater and surface water at the sub-catchment scale, a transient groundwater model with an upscaling procedure combining remote sensing data and field observations for the Bulang sub-catchment was constructed. The catchment water balance was analysed by considering vegetation types with the Normalized Difference Vegetation Index (NDVI), determining evaporation rates by combining sap flow measurements and NDVI values, recorded precipitation, measured river discharge and groundwater levels from November 2010 to October 2011. A comparison between a simple water balance computation, a steady state groundwater flow model, and the transient groundwater flow model indicated that different land use scenarios would result in different river discharges. It was shown that 91% of the precipitation was consumed by the crops, bushes and trees; only 9% of the annual precipitation became net groundwater recharge, which maintained a stable stream discharge during the year with observations. Four land use scenarios were analysed as (1) the quasi natural state of the vegetation covered by desert bushes; (2) the current land use/vegetation types; (3) the change of crop types to dry resistant crops; and (4) the ideal land use covered by dry resistant crops and desert bushes. These four scenarios were simulated and compared with the measured data from 2011, which was a dry year. Furthermore, scenarios (2) and (4) were evaluated under normal and wet conditions for the years 2009 and 2014, respectively. The simulation results show that dry resistant bushes and certain crops can significantly increase net groundwater recharge, which leads to an increase of groundwater storage and river discharges. The depleted groundwater storage during the dry year could be restored during the normal and wet years such that groundwater provides a reliable resource to sustain river discharge and the dependent vegetation in the area.

The fourth topic presents an investigation of human activities on the interactions between groundwater and surface water in the Hailiutu catchment. The isotopic and chemical analysis of surface water and groundwater samples identified groundwater discharges (seepage?) to the river along the Hailiutu River. Mass balance equations with chemical profiles were used to estimate the spatial distribution of groundwater seepage rates along the river. The temporal variations of groundwater and surface water interactions were investigated using the hydrograph separation method. A numerical groundwater model was constructed to simulate groundwater discharges along the river and analyse the effects of water use in the catchment. The simulated seepage rates along the river compare reasonably well with the seepage estimates from the chemical profile measurements in 2012. The impacts of human activities including river water diversion and groundwater abstraction, on river discharge were analysed by calculating the differences between the simulated natural groundwater discharge and the measured river discharge. The water use in the Hailiutu River increased from 1986 to 1991, reached its highest level from 1992 to 2000, and decreased from 2001 onwards. The reduction of river discharge might have had negative impacts on the riparian ecosystem and the water availability for downstream users. Thus, the interactions between groundwater and surface water as well as the consequences of human impacts should be taken into account when implementing sustainable water resources management.

This research has addressed multi-disciplinary topics on hydrology, climate change, land use change and upscaling methods in terms of interactions between groundwater and surface water in the Hailiutu River catchment, Erdos plateau, China. The river flow regime has been intensively influenced by human activities, such as the construction of reservoirs, water diversion, and groundwater exploitation. Since groundwater discharge dominates the overall discharge in the Hailiutu River catchment, the replacement of the current vegetation with less water consuming crops and bushes would increase the groundwater discharge to the river. Hence, the optimized land use would benefit water resources management as well as the ecosystem in the Hailiutu River catchment. The main findings in this study provided valuable insights for the scientific community and sustainable policy making.

Samenvatting

Eeuwenlang zijn het grondwater en oppervlaktewater beschouwd als geïsoleerde onderdelen van de hydrologische cyclus bij de toepassing van waterbeheer. Grondwater heeft echter hydraulische connecties met het oppervlaktewater en zij vormen samen een enkel systeem. De interacties tussen het grondwater en oppervlaktewater omvat complexe processen op vele verschillende temporele en ruimtelijke schaalgroottes. Een groeiend aantal studies over de interacties tussen grondwater en oppervlaktewater zijn uitgevoerd in berg en laagland gebieden, zowel in tropische als gematigde zones met verschillende klimatologische omstandigheden. Menselijke activiteiten hebben de natuurlijke connecties tussen grondwater en oppervlaktewater beinvloed. De dynamiek van de interacties tussen het grondwater en oppervlaktewater, evenals de reactie op menselijke verstoringen, wekt onze interesse naar mogelijke effecten.

Het stroomgebied van de Hailiutu rivier dat zich bevindt op het semi-aride Erdos plateau in China, heeft te lijden onder conflicten die verband houden met de exploitatie van watervoorraden voor sociale en economische ontwikkeling aan de ene kant en de bescherming van het ecosysteem anderzijds. In de laatste decennia heeft de lokale gemeenschap te maken gehad met tekorten aan water, dalende grondwaterspiegels, en verslechtering van het ecosysteem. Hoewel grondwater een zeer belangrijke rol in de watervoorziening voor de lokale gemeenschap speelt en het grondwater-afhankelijke ecosysteem in het stroomgebied van de Hailiutu rivier uitermate delicaat is, zijn de interacties tussen grondwater en oppervlaktewater, alsmede de impact die de mens heeft op de hydrologische processen, niet goed onderzocht. Het verkrijgen van inzicht in de mechanismen die de interacties tussen het grondwater en oppervlaktewater aansturen, is van cruciaal belang voor het bereiken van duurzame ontwikkeling.

Deze thesis presenteert een systematische aanpak voor het onderzoeken van de interacties tussen grondwater en oppervlaktewater, bestaande uit een statistische analyse van de dynamiek van water voorkomens op stroomgebiedsniveau, de toepassing van meerdere in-situ methoden voor het uitvoeren van observaties die bijdrages van grondwater op een lokale schaal vaststellen, het uitvoeren van eenvoudige tot complexe analyses op basis van de waterbalansen van sub-stroomgebieden, en de kwantificering van temporele en ruimtelijke interacties tussen grondwater en oppervlaktewater voor het hele stroomgebied van de Hailiutu rivier met gebruik making van 'tracer' en modellerings technieken.

Op grond van de uitgevoerde systematische studie naar de de kwantitatieve beoordeling van de interacties tussen grondwater en oppervlaktewater in het stroomgebied van de Hailiutu rivier, kunnen inhoud en conclusies van het proefschrift als volgt worden samengevat.

Ten eerste wordt het belang van interacties tussen grondwater en oppervlaktewater, alsmede de menselijke invloed daarop en de gevolgen voor duurzame ontwikkeling en bescherming van het ecosysteem van het Erdos plateau, benadrukt. Informatie over de algemene achtergrond van het Erdos plateau en het beoogde stroomgebied van de Hailiutu rivier worden geintroduceerd. Vervolgens krijgen de onderzoeksdoelstellingen en de gevolgde methodieken de volle aandacht.

De veranderingen in het afvoer regime van de Hailiutu rivier zijn systematisch geanalyseerd om de effecten van klimaat en menselijke activiteiten vast te stellen. Met statische methodes konden significante veranderingen in het afvoer regime worden bepaald. Dit heeft geresulteerd in het vaststellen van vijf perioden met een typisch afvoer regime aan de hand van vier veranderingen in het afvoer verloop die plaatsvonden in 1968, 1986, 1992 en 2001. Het afvoer regime weerspiegelt de quasi-natuurlijke situatie van hoge variabiliteit en grote amplitudes van de 6 maandelijkse periodieke fluctuaties gedurende de periode van 1957 tot 1967. De hoge afvoeren van de rivier werden afgeroomd in de periode van 1968 tot 1985 door de bouw van twee reservoirs. In de periode van 1986 tot 1991 daalde de afvoer van de rivier verder ten gevolge van de gecombineerde invloed van het aftappen van water uit de rivier en de toename van grondwateronttrekkingen voor irrigatie. In de periode van 1992 tot 2000 was de afvoer van de rivier minimaal zowel met betrekking tot volume als variabiliteit. Deze periode correspondeert met een grote uitbreiding van de landbouwgebieden op het plateau. De afvoeren van de rivier herstelden zich in de periode van 2001tot 2007 toen regeringsbeleid werd ingevoerd om landbouwgrond te converteren naar natuurlijk bos en grasland. Er kon vastgesteld worden dat klimatologische elementen zoals neerslag en lucht temperatuur maar een zeer klein effect hebben gehad op de veranderingen in het afvoer regime. Het implementeren van hydraulische werken voor de berging en het aftappen van oppervlaktewater, grondwateronttrekkingen, en de aanpassingen in het landgebruik waren

X

verantwoordelijk voor de veranderingen van het afvoer regime in het stroomgebied van de Hailiutu rivier.

Het tweede aandachtspunt vormen de in-situ methoden voor het uitvoeren van observaties op lokale schaal. De identificatie en kwantificering van de interacties tussen het grondwater en oppervlaktewater werden uitgevoerd langs de Bulang rivier die een zijrivier is van de Hailiutu rivier. Er is gebruik gemaakt van hydraulische, temperatuur, hydro-chemische en isotopen methodes waarbij de veldmetingen over een periode van een jaar zijn uitgevoerd. Volgens de metingen domineert de grondwater afvoer naar de rivier de hydrologische processen. De waargenomen grondwaterstanden waren altijd hoger dan de waterstand van de rivier en dit indiceert de afvoer van grondwater naar de rivier. Temperatuurmetingen van het rivier water, in het sediment bed van de rivier op verschillende dieptes, en van het grondwater zelf bevestigden de opwaartse stroom van grondwater naar de rivier gedurende alle seizoenen. De resultaten van het onderverdelen van de afvoer hydrograaf op basis van 'tracer' metingen onthulden dat, zelfs tijdens zware regenval, het grondwater meer bijdraagt aan de toegenomen afvoer van de rivier dan het afstromende oppervlaktewater. Bovendien werd de grondwater afvoer langs een geselecteerd lengteprofiel van de rivier gekwantificeerd door afvoermetingen uit te voeren en die te combineren met geleidbaarheid (EC) metingen in het rivier water, onder natuurlijke omstandigheden en na een constante injectie van een 'tracer'.

Om de temporele en ruimtelijke interacties tussen grondwater en oppervlaktewater op de schaalgrootte van een sub- stroomgebied te bepalen, werd een tijdsafhankelijk grondwater model geconstrueerd, voorzien van een opschalingsprocedure die gebruik maakt van tele-detectie gegevens en veldwaarnemingen uit het sub- stroomgebied van de Bulang rivier. De waterbalans van het gebied werd opgemaakt door vegetatietypes te classificeren met de Genormaliseerde Verschil Vegetatie Index (NDVI) methode, en verdampingssnelheden te bepalen door het combineren van 'sap' stroom metingen en NDVI-waarden. Daarnaast werden ook de gemeten neerslag en afvoer van de rivier, alsmede de grondwaterstanden vanaf november 2010 tot en met oktober 2011 in overweging genomen. Door de eenvoudige waterbalans, een tijdsonafhankelijk grondwater model, en het tijdsafhankelijke model met elkaar te vergelijken, kon de conclusie worden getrokken dat verschillen in landgebruik leiden tot verschillende groottes van de afvoer in de rivier. Voor het observatie jaar bleek dat 91% van de neerslag werd verbruikt door de gewassen, struiken en bomen. Slechts 9% van de jaarlijkse neerslag werd omgezet in grondwateraanvulling, die vervolgens een stabiele afvoer van de rivier garandeerde. Tot slot werden de 4 volgende scenario's voor landgebruik geanalyseerd (1) het landgebruik reflecteert de quasi natuurlijke toestand van een vegetatie dek bestaande uit woestijn struiken; (2) het landgebruik en de vegetatie en gewas typen zien eruit zoals momenteel gangbaar is; (3) de huidige landbouw gewassen worden vervangen door droogte-resistente gewassen; en (4) het ideale landgebruik bestaande uit woestijn struiken en droogte resistente gewassen wordt ingevoerd. Deze vier scenario's werden gesimuleerd met de modellen en vergeleken met meetgegevens uit 2011 dat een droog jaar was. Bovendien werden scenario's (2) en (4) geëvalueerd onder normale en natte omstandigheden waarvoor respectievelijk de jaren 2009 en 2014 als uitgangspunt dienden. De simulatie resultaten laten zien dat woestijn struiken en droogte-resistente gewassen de grondwateraanvulling aanzienlijk verhogen, hetgeen weer leidt tot een toename van de grondwater berging en afvoer in de rivier. De sterk verminderde grondwater berging ontstaan tijdens het droge jaar kon worden hersteld tijdens het normale en natte jaar in die mate dat grondwater een betrouwbare bron van water bleek om de afvoer in de rivier en de grondwater-afhankelijke vegetatie in het gebied te ondersteunen.

Het laatste aandachtspunt behelst het onderzoek naar de effecten van menselijke activiteiten op de interactie tussen het grondwater en oppervlaktewater in het stroomgebied van de Hailiutu. Isotopen en chemische analyses van grondwater en oppervlaktewater monsters toonden wederom de afvoer van grondwater naar de rivier aan. Massabalans vergelijkingen samengesteld op basis van chemische profielen zijn gebruikt om een inschatting te maken van de ruimtelijke spreiding van de afvoer van grondwater langs de loop van rivier. Temporele variaties in de interacties van grondwater en oppervlaktewater werden onderzocht door gebruik te maken van methodes voor het onderverdelen van de afvoer hydrograaf van de rivier. Grondwater modellering werd ingezet om langs de loop van de rivier de afvoer van grondwater te simuleren en het effect van menselijk water gebruik in het stroomgebied te analyseren. De gesimuleerde afvoeren van grondwater naar de rivier vertonen een redelijke overeenkomst met de geschatte afvoeren gemaakt op basis van de gemeten chemische profielen in 2012. De impact van menselijke activiteiten zoals het aftappen van water uit de rivier of het implementeren van grondwateronttrekkingen, op de afvoer van de rivier werd geanalyseerd door het berekenen van de verschillen tussen de gesimuleerde natuurlijke afvoer van grondwater naar de rivier en de gemeten water afvoer in de rivier.

Het aftappen van water uit de Hailiutu rivier nam toe van 1986 tot 1991, bereikte een hoogste niveau van 1992 tot 2000 en daalde daarna vanaf 2001. De vermindering van de afvoer van de rivier zou negatieve gevolgen kunnen hebben gehad voor het ecosysteem langs de rivier en de beschikbaarheid van water voor benedenstroomse gebruikers. Met de interacties tussen grondwater en oppervlaktewater, alsmede de gevolgen van menselijke activiteiten moet dus rekening worden gehouden bij de uitvoering van duurzaam waterbeheer.

Dit onderzoek heeft zich geconcentreerd op de inzet van multidisciplinaire onderwerpen op het gebied van de hydrologie, klimaatverandering, landgebruik en opschalingsmethodes om de interacties tussen grondwater en oppervlaktewater te analyseren in het stroomgebied van de Hailiutu rivier op het Erdos plateau in China. Het afvoer regime van de rivier werd ernstig beïnvloed door menselijke activiteiten, zoals de aanleg van stuwmeren, het aftappen van water en het implementeren van grondwateronttrekkingen. Aangezien de afvoer van grondwater de algemene water afvoer in het stroomgebied van de Hailiutu rivier domineert, zou de vervanging van de huidige vegetatie door droogte-resistente gewassen en struiken, de afvoeren doen vergroten. Een geoptimaliseerd landgebruik zou het waterbeheer positief beinvloeden, alsmede het ecosysteem in het stroomgebied van de Hailiutu rivier verbeteren. De belangrijkste bevindingen van deze studie leveren waardevolle inzichten op voor de wetenschappelijke gemeenschap en duurzame beleidsvorming.

Acknowledgements

It was a big challenge to do the annual shift between scientific research at UNESCO-IHE in The Netherlands and practical environmental impact assessment consultancy in China during the last 7 years PhD research. They were full of challenges along with opportunities. Challenges in setting-up the field research site, analysing measurements, and writing scientific articles forced me to develop as scientific researcher. Opportunities in working with renowned experts in the field and meeting fellow students from all over the world were most valuable experiences in this study. I am grateful to those who have contributed to the completion of this thesis.

I am very grateful to my wife Zhou Lingyun, my draught Yang Jingyun, my father and mother, and my parents in law, all of you always support me with the encouragement, and take care of family during my absent.

I would like to express my sincere and deepest appreciations to my co-promoter Dr. Yangxiao Zhou, I could not have finished this PhD thesis without his continuous, devoted, and careful guidance. Your innovative, critical, and inspiriting supervision are the most reliable and dependable source from the beginning of the research and to my future career.

Many thanks are due to my supervisor Dr. Jochen Wenninger for all help in the field investigation, instrument installation, data analysis at laboratory, and constructive discussions in the study.

My extreme gratitude goes to my promoter Prof. Stefan Uhlenbrook for his helpful, comfortable, and professional support. His wise advice, critical and creative insights, and wealth of broad knowledge had tremendously effects on my study.

I would like to thank Prof. Wan Li, Prof. Wang Xusheng, Prof. Jin Xiaomei, Dr. Chen Jinsong, Prof. Hu Fusheng, Dr. Hou Lizhu, among many others at China University of Geosciences (Beijing) for their support and cooperation. The support from Prof. Xie Yuebo, Prof. Zhang Danrong, Prof. Jiang Cuiling at Hohai University are acknowledged.

Special thanks go to Erdos research team members (Hou Guangcai, Yin Lihe, Huang Jinting, Dong Ying, Chang Liang, Wang Xiaoyong, Dong Jiaqiu) at Xi'an Center of Geological Survey. Without their cooperation and support, I could not operate field experimental facilities at the Hailiutu catchment.

Many friends in the Netherlands and China are appreciated for enjoying happy and relaxing times together. They are Li Shengyang, Wang Chunqin, Wan Yuanyang, Guo Leicheng, Yan Kun, Zhu Xuan, Ye Qinghua, Wan Taoping, Xu Zhen, Chen Qiuhan, Chen Hui, Pan Quan, Xu Ming, Wang Wen, Li Hong, Zhang Guoping, Zuo Liqin, Wang Hao, etc. Thank you all for your companion during my stay in Delft.

Finally, financial support of the study is acknowledged: China Scholarship Council, Netherlands Asian facility for China programme, Honor Power Foundation, and IHE Internal Research Fund.

YANG Zhi

Anhui, China

November, 2017

Contents

Figures and Tables

List of figures

chemical profile and measured river discharges at Hanjiamao hydrological station and simulated by groundwater flow model along the Hailiutu River................................109

List of tables

1. Introduction

1.1 Background

Groundwater and surface water have been considered as isolated components of the hydrological cycle for centuries, but they interact in a variety of ways depending on the physiographic settings (e.g. Winter, 1999, Sophocleous, 2002). Groundwater, as well as surface water resources, are basic conditions for social economic developments worldwide. However, interactions between groundwater and surface water are difficult to measure and quantify (Winter, 1999), which have been considered separately in water resource management and policy formulations. Interactions between groundwater and surface water play very important roles for stream ecosystems (Findlay, 1995) and hence have consequence for ecology, river restoration and conservation (Boulton, et al., 2010) and the protection of groundwater-dependent ecosystems (Zhou, et al., 2013, Bertrand, et al., 2014). However, the relationships between the groundwater and the surface water bodies, such as rivers, lakes, reservoirs, wetlands, have not been fully understood. Climate and human activities were considered to be the two main influential factors by hydrologists and hydrogeologists (Milliman, et al., 2008, Uhlenbrook, 2009, Zhao, et al., 2009, Xu, 2011), but the distinction for them on specific cases remains difficult and debatable.

The recent studies on the interactions between groundwater and surface water utilized multiple methods, including direct field investigations (e.g. Oxtobee and Novakowski, 2002) and measurements (Anderson, et al., 2005) to determine differences in hydraulic heads, chemical and isotopic tracers (e.g. Wenninger, et al., 2008), temperature studies (Conant, 2004, Schmidt, et al., 2007). Remote sensing combined with field observations have been widely employed in hydrological studies by means of upscaling procedures (e.g. Ford, et al., 2007). Due to the temporal and spatial flexibility, numerical modelling approaches have been carried out to study the interactions between groundwater and surface water for transition zone water (Urbano, et al., 2006), the importance of water balance in a mesoscale lowland river catchment (Krause and Bronstert, 2007, Krause, et al., 2007), small catchments (Jones, et al., 2008, Guay, et al., 2013) with different scenarios (Gauthier, et al., 2009), to conduct water resources assessment (Henriksen, et al., 2008), to evaluate the impacts of best agricultural management practices (Cho, et al., 2010), and to determine the impacts of climate changes (Scibek, et al., 2007). Apart from natural processes, the anthropogenic effects cannot be neglected in an investigation on the spatial and temporal relationships between groundwater and surface water. However, the quantification of human impacts on the interactions between groundwater and surface water would only be achieved by multiple methods according to previous studies. This thesis is formulated based on summarizing those scientific findings on identification and quantification of interactions between groundwater and surface water.

Integrated water resources management has been implemented in many areas in China with the goal of achieving sustainable development (Ministry of Water Resources, China, 2005). Nevertheless, there are both institutional and technical difficulties that exist when people are facing water shortage, especially for those who live in the arid/semi-arid regions where the groundwater plays a very important role in the water supply for the society and ecosystem. The interactions between groundwater and surface water are the most important part of water cycle in arid and semi-arid region, where the environment is very sensitive to water resources development. Previous studies have shown that water resource scarcity is one of limiting factors for socio-economic development in northwest China. Water management has focused on surface water and groundwater as separate resources for decades in China. The local authorities of Erdos City are concerned more about the ecosystems, sustainable development and river basin management in recent years. However, water management has focused on surface water or groundwater as separate resources. Furthermore, groundwater requirements for ecosystems are not well understood and are often neglected or poorly managed. These interactions can have significant implications for both water quantity and quality (Bordie, et al., 2007). Thus, the understanding of the interactions between the groundwater and surface water on the Erdos plateau becomes crucial to water resources management in terms of both water quantity and quality.

The Erdos plateau administratively belongs to Erdos City of Inner Mongolia and the Yulin City of Shaanxi province with an area of approximately 200,000 km^2. Half of the area is covered by deserts and bare rocks. The total population is approximately 26 million. The climate is typical inland arid to semi-arid. The precipitation is scarce, with annual average varying from 400 mm/year in the east to 200 mm/year in the west. Potential evaporation is very high, ranging from 2000 to 3500 mm/year. Since the potential evaporation is larger than the precipitation, the surface water resources are limited. The main water resource in the plateau is groundwater (Gao, et al., 2004). Terrestrial ecosystems depend mainly on groundwater. However, overgrazing and cultivation activities accelerated the deterioration of the ecosystem. Desertification, soil loss and land degradation are major problems in the Erdos plateau (Wang, 2008). Located in the middle catchment of the Yellow River system in the centre of China, Erdos is one of new energy bases for China and is targeted as a priority area of western development strategy for China in the 21st century. Exploitation of coal, natural gas, oil and mineral resources has sped up socio-economic development in the region. The challenge for local governments is to achieve sustainable water resources development to meet increasing water demands from the industry, agriculture, society and ecosystem. This has already overstressed scarce water resources and may have disastrous consequences on fragile ecosystems. Groundwater on the Erdos plateau is historically treated as loss term or static storage in surface water-oriented water balance management. The neglect of groundwater and surface water interactions has caused depletion of stream flows, aquifers and degradation of groundwater-dependent ecosystems. It is increasingly recognized that groundwater and surface water interactions occur in different forms under different physiographic and hydro-climatic conditions. However, these interactions are often altered by anthropogenic interventions, to an extend that is largely unknown.

The Hailiutu River catchment has been selected for detailed research on interactions between groundwater and surface water by the UNESCO-IHE, the Netherlands, China University of Geosciences (Beijing) and Hohai University, China, for its typical characters in terms of the interactions between groundwater and surface water, demands of integrated water resources management as well as the influence of human activities on hydrological process. The river discharge has significantly decreased according to a preliminary analysis. A multi-disciplinary approach research has been carried out to quantify spatial and temporal interactions between groundwater and surface water in the Hailiutu River catchment and Bulang sub-catchment. The results should provide a solid scientific basis for an integrated approach for water resources management in river basins, which will hopefully lead to a reversal of the trend of desertification and ecosystem degradation in the Erdos Plateau.

Surface water is connected with groundwater in the Erdos Plateau in many different ways, which leads to high complexity of the hydrological processes in the region. The improved understanding and knowledge of the hydrological processes in interactions between groundwater and surface water will facilitate efficient conjunctive water resources management. A healthy natural environment with sustainable water resources development is a primary condition for sustainable livelihoods of the local population and sustained economic growth and poverty alleviation. Although the interactions between the groundwater and surface are crucial, studies on the relationships between the groundwater abstraction and the discharge in the streams are very limited in this region due to the complexity of the groundwater and surface water systems. Therefore, a better understanding of the interaction between groundwater and surface water is needed for the sustainable and integrated water resources management. This research aims to extend our knowledge of the exchange between groundwater and surface water on Erdos Plateau in order to provide reliable scientific information to decision-makers for the conjunctive sustainable water resources management.

1.2 Research objectives and approach

The aim of this study is to understand the underlying processes and to quantify the interactions between groundwater and surface water in the Erdos Plateau.

The specific objectives are:

1) To explore the mechanism of water exchanges between groundwater and surface water at reach scale;

2) To determine the dominant processes and spatial-temporal variations of

groundwater and surface water interactions in the selected sub-catchment;

3) To provide useful inputs towards a conjunctive groundwater-surface water development plan to meet the water demand for socio-economic development while allocating environmental flows for groundwater dependent ecosystems;

4) To understand historical variations of groundwater levels with the help of a constructed physically-based coupled groundwater and surface water model;

5) To identify and quantify the spatial and temporal variability of interactions between groundwater and surface water in the Hailiutu catchment scale;

6) To estimate the effects of groundwater extraction on the river baseflow and dependent ecosystem in the connected river-aquifer-vegetation system in the Hailiutu catchment;   and

7) To distinguish the impacts of different Human activities from climate variety on the interactions between groundwater and surface water at Hailiutu catchment.

To achieve the above objectives, an integrated approach with combination of statistical analysis of historical data, field observation, temperature measurements, chemical and isotopic analysis on groundwater and surface water samples, and numerical modelling have been developed and conducted in the selected Hailiutu catchment and it's tributary Bulang River. After distinguishing the impacts from the climate variables and human activities by a statistical analysis on the historical hydro-meteorological observations, the combined methods with multiple upscaling approaches for investigating interactions between groundwater and surface water in the Hailiutu catchment were applied at the reach and catchment scales. The temporal and spatial variations of interactions between groundwater and surface water have been investigated at the reach scale in the Bulang tributary, sub-catchment, and basin scale. Furthermore, the identification and quantification of components of hydrological processes in the Hailiutu catchment, as well as the impacts of climate and human activities on the interactions between groundwater and surface water, were conducted by applying multiple methods. The research framework is presented in Figure 1.1.

1.3 Innovation and challenge

The hydrological and geomorphological analysis in the semi-arid Hailiutu catchment in Erdos Plateau indicate that the groundwater plays very important role not only in sustaining the hydrological cycle, but also for the ecosystem and society. Recent researches focused more on floods and on the reduction of sediment load in the Yellow river caused by soil and water conservation measurements in the Loess Plateau, and the impacts of climate change on the hydrological processes. However, the quantification of interactions between groundwater and surface water for water resources management has not been studied.

This thesis presents a first study on the hydrological alterations of stream flow by different human activities in Erdos Plateau, which was ascribed to the climate change and soil and water conservation measurements in the previous studies in the middle Yellow river.

First, multiple methods have been applied in this study to determine groundwater-surface water interactions. Considering the limitations of individual field measurement for determining interactions between groundwater and surface water, this thesis adopted multiple methods for identifying and quantifying interactions between groundwater and surface water that consists of hydraulic, temperature, hydrochemistry, isotope and numerical modelling methods. The quantity and quality of the groundwater and surface water can be simultaneously or independently affected by solute exchange among soil, rock, and water. The salt accumulation at top soil due to the evaporation and artificial solute release will also increase the complexity of the chemical components. Hence, the method of estimation of groundwater seepage rates using the chemical profile along the stream in this study provides an innovative efficient method to directly measure the seepage rates along the river with low cost and high reliability. Second, this thesis presents the systematic analysis for the quantification of the interactions between groundwater and surface water at reach, sub-catchment, and catchment scales through different models. Third, the impacts of human activities such as the construction of the reservoirs, water diversion in the flood plain, groundwater exploitation at basin scale and the land use at sub-catchment scale, on the connectivity between groundwater and surface water were analysed.

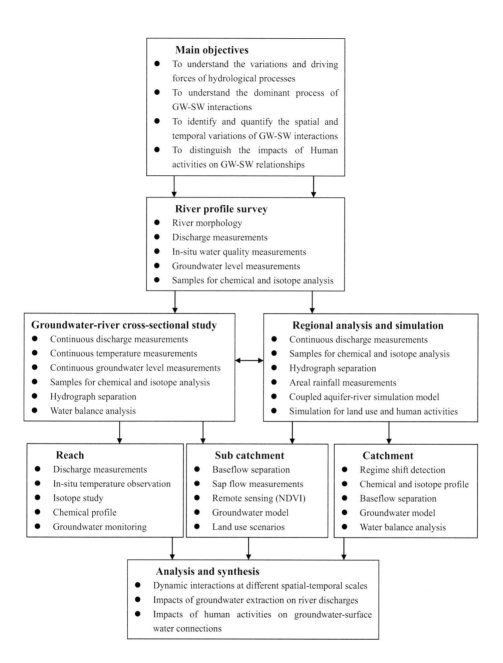

Figure 1.1 Methodologies framework in this study.

Among the scientific challenges in this study, the most significant obstacle is reducing the uncertainties in determining evaporation rate and groundwater recharge at catchment scale during upscaling procedures. Remote sensing and a geographic information system (GIS) facilitated estimation of areal evaporation rates from individual experimental observations with Normalized Difference Vegetation Index (NDVI) in this thesis, which provided a novel comprehensive method for hydrological studies at the catchment and sub-catchment scales. Furthermore, conflicts between water use by human activities and the groundwater dependent ecosystem were tackled by scenario analysis of land use changes.

In addition to the scientific challenges on the determination of spatial-temporal variations of groundwater and surface water interactions in the catchment, the impacts of human activities on the available water resources are critical influential facts that need to be quantified. For management challenges, the mutual benefits for ecosystem conservation and water resource use for social and economic developments should be achieved by means of management of groundwater abstractions.

The results of this PhD study provide valuable references for sustaining conjunctive water resources management in arid and semi-arid regions. The scientific findings during this study also benefit future anti-desertification and ecosystem restoration projects via the land use reformation. The goal of sustainable development could be achieved based on the results of interactions between groundwater and surface water in this thesis.

1.4 Thesis outline

This PhD thesis is formulated based on four peer-reviewed international journal papers. These papers are organized in this thesis as separate chapters. The detailed structure of this thesis is as follows:

Chapter 1 identifies the problems and formulates research objectives.

Chapter 2 addresses general aspects of shifts in the flow regime in the Hailiutu River and possible driving forces (Human activities vs Climate change) in terms of statistical analysis on the characteristics of river discharge, climatic variables, and the crop area. Human interference has significant impacts on the hydrological processes, especially on the interactions between groundwater and surface water.

Chapter 3 investigates the interactions between groundwater and surface water in the Bulang tributary of the Hailiutu River by means of hydraulic, temperature, hydrochemistry, and isotopic methods, which indicates the groundwater discharges to the rive dominate the hydrological processes in the Bulang tributary. Furthermore, a quantitative estimation of groundwater discharge along a gaining reach with chemical profile and discharge measurements was proposed.

Chapter 4 evaluates the performance of groundwater model with remote sensing and field observation data of actual evaporation for bushes and maize on groundwater-surface water interactions under different land use scenarios within the Bulang sub-catchment. The recommended land use with less water consumption benefits for a shallow groundwater depth and stable river discharge.

Chapter 5 quantifies the spatial variation of human impacts on the groundwater and surface water interactions in the Hailiutu River by comparing simulated groundwater discharges with hydrological observations and calculated groundwater discharge rates with chemical profile measurements along the Hailiutu River.

Chapter 6 draws the syntheses of the previous chapters, the main conclusions, recommendations, and future research areas.

2. The flow regime shift in Hailiutu river[*]

Abstract: Identifying the causes (climate vs. human activities) for hydrological variability is a major challenge in hydrology. This paper examines the flow regime shifts, changes in the climatic variables such as precipitation, evaporation, temperature, and crop area in the semi-arid Hailiutu catchment in the middle section of the Yellow River by performing several statistical analyses. The Pettit test, cumulative sum charts (CUSUM), regime shift index (RSI) method, and harmonic analysis were carried out on annual, monthly, and daily discharges. Four major shifts in the flow regime have been detected in 1968, 1986, 1992 and 2001. Characteristics of the flow regime were analyzed in the five periods: 1957-1967, 1968-1985, 1986-1991, 1992-2000, and 2001-2007. From 1957 to 1967, the flow regime reflects quasi natural conditions of the high variability and larger amplitude of 6 months periodic fluctuations. The river peak flow was reduced by the construction of two reservoirs in the period 1968-1985. In the period of 1986-1991, the river discharge further decreased due to the combined influence of river diversions and increase of groundwater extractions for irrigation. In the fourth period of 1992-2000, the river discharge reached lowest flow and variation in corresponding to a large increase in crop area. The flow regime recovered, but not yet to natural status in the fifth period of 2001-2007. Climatic factors are found not likely responsible for the changes in the flow regime, but the changes in the flow regime are corresponding well to historical land use policy changes.

2.1 Introduction

The temporal pattern of river flow over a period of time is the river flow regime, which is a crucial factor sustaining the aquatic and riverine ecosystems. Regime shifts are defined in ecology as rapid reorganizations of ecosystems from one relatively stable state to another (Rodionov and Overland, 2005). Flow regime shifts represent

---

[*].This chapter is based on paper The causes of flow regime shifts in the semi-arid Hailiutu River, Northwest China. Yang, Z., Zhou, Y., Wenninger, J., & Uhlenbrook, S. (2012). Hydrology and Earth System Sciences, 16, 87–103, DOI 10.5194/hess-16-87-2012.

relatively sudden changes in temporal characteristics of river discharges in different periods. It is widely accepted that climate change and human activities are the main driving forces for hydrological variability (Milliman, et al., 2008, Zhao, et al., 2009, Xu, 2011). However, distinguishing the causes for the flow regime shifts is still a major challenge in hydrology.Studies show that flow regime shifts in river basins can be ascribed to the changes in climatic variables, land cover and land use, river regulations, and other human activities; for example, soil and water conservation measures. The climatic variables were considered as the major driving factors for long-term changes in river discharge (Arnell and Reynard, 1996, Neff, et al., 2000, Middelkoop, et al., 2001, Christensen, et al., 2004, Jha, et al., 2004, Wolfe, et al., 2008, Timilsena, et al., 2009, Masih, et al., 2011). The impacts of future climate changes on stream discharge were also predicted (Gellens and Roulin, 1998, Chiew and McMahon, 2002, Eckhardt and Ulbrich, 2003, Drogue, et al., 2004, Thodsen, 2007, Steele-Dunne, et al., 2008). The changes in land cover (Matheussen, et al., 2000, Cognard-Plancq, et al., 2001, Costa, et al., 2003, Bewket and Sterk, 2005, Poff, et al., 2006, Guo, et al., 2008) and land use (Fohrer, et al., 2001, Tu, 2006, Zhang and Schilling, 2006, Rientjes, et al., 2010, Masih, et al., 2011) would eventually alter the river discharge by influencing the runoff generation and infiltration processes. The construction of dams can significantly reduce the high flows and increase the low flows (Maheshwari, et al., 1995, Magilligan and Nislow, 2005). The hydrological response also depends on a combination of precipitation, evaporation, transpiration, basin permeability and basin steepness (Lavers, et al., 2010) or runoff generation in headwater catchments, impoundments in small dams and increased extractions for irrigated crop production (Love, et al., 2010). These studies mainly focused on the relationship between the mean annual stream flow and the corresponding factors by performing statistic tests on indicators of hydrological alterations or comparing modelled and measured discharges.

In China, the relations between the stream flow, precipitation and temperature were investigated in the Tarim River (Chen, et al., 2006), Yellow River (Fu, et al., 2007, Hu, et al., 2011), Wuding River (Yang, et al., 2005) and Lijiang River (He, et al., 2010). Zhao et al. (2009) studied the streamflow response to climate variability and human activities in the upper Yellow River Basin, and suggested that the climate effects accounted for about 50% of total streamflow changes while effects of human activities on streamflow accounted for about 40%. But the type of human activities was not identified. Furthermore, the changes in river discharge induced by soil and water conservation measures were examined in the Loess Plateau (Li, et al., 2007, Dou, et al., 2009) and Wuding River (Xu, 2011). The effects of dam construction (Yang, et al., 2008) and operation (Yan, et al., 2010) on flow regimes in the lower Yellow River were assessed by analyzing the indicators of hydrological alterations (Richter, et al., 1996), which suggested that dams affect the stream flow by increasing the low flows and decreasing the high flows.

Much of the present studies focus on the relations among the changes in climate and their linkage with the streamflow regime. However, the regime shift of the river discharge can also be caused by the human activities, but very often these factors cannot be distinguished (Uhlenbrook, 2009). Although some studies on climate change, dam regulation, human activities of soil and water conservations, and their effects on the river discharge have been conducted in the Loess Plateau of the Yellow River and its tributaries, no study focused so far on the regime shifts caused by human activities vs. climate controls in the sandy region in the middle section of the Yellow River Basin.

This paper reveals the flow regime shifts by means of detecting changes in annual, monthly, and daily characteristics of the river discharge and connects with changes in climate, water resources development, and land use in the sandy region of the middle section of the Yellow River Basin. The results provide a better understanding of the hydrological response to climate and human activities in a semi-arid area.

2.2 Material and methods

2.2.1 Study area

The Hailiutu catchment is located in the middle section of the Yellow River Basin in Northwest China. The Hailiutu River is one of the branches of the Wuding River, which is the major tributary of the middle Yellow River (Figure. 2.1). The total area of the Hailiutu catchment is around 2645 km^2. The surface elevation of the Hailiutu catchment ranges from 1020 m in the southeast to 1480 m above mean sea level in the northwest. The land surface is characterized by undulating sand dunes, low hills at the northern and western water divide, and an U-shaped river valley in the downstream area. A hydrological station is located at the outlet of the Hailiutu catchment near Hanjiamao village with a mean annual discharge of 2.64 m^3/s for the period

1957–2007. There is only one tributary of the Hailiutu River, named Bulang River, which is situated at the middle part of the catchment. There are two reservoirs constructed; one at the upstream of the Hailiutu River and the other one at the Bulang tributary for local water supply. The information on the construction of the reservoirs and water diversions is listed in the Table 2.1.

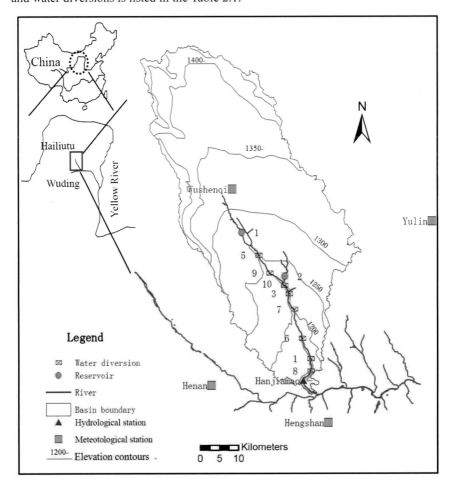

Figure 2.1 Map of the Hailiutu catchment, the numbers nearby the stations are indices of hydraulic engineering works in the Table 2.1.

Table 2.1 Hydraulic engineering works in the Hailiutu catchment.

No.	Name	Year of construction	Elevation m (a.m.s.l)	Type	Water use
1	Chaicaoba	1970	1072	Diversion dam	Irrigation for 53 ha crop land
2	Tuanjie	1971	1218	Reservoir	Water supply for power plant; Irrigation for 33 ha crop land

14

No.	Name	Year of construction	Elevation m (a.m.s.l)	Type	Water use
3	Maluwan	1972	1124	Diversion dam	Irrigation for 187 ha crop land
4	Geliugou	1972	1166	Reservoir	Irrigation for 33 ha crop land
5	Caojiamao	1989	1184	Diversion dam	Irrigation for 93 ha crop land
6	Hongshijiao	1992	1082	Diversion dam	Irrigation for 113 ha crop land
7	Shuanghong	1995	1101	Diversion dam	Irrigation for 100 ha crop land
8	Wanjialiandu	1995	1043	Diversion dam	Irrigation for 133 ha crop land
9	Wujiafang	1997	1150	Diversion dam	Irrigation for 60 ha crop land
10	Weijiamao	2008	1130	Diversion dam	Irrigation for 67 ha crop land

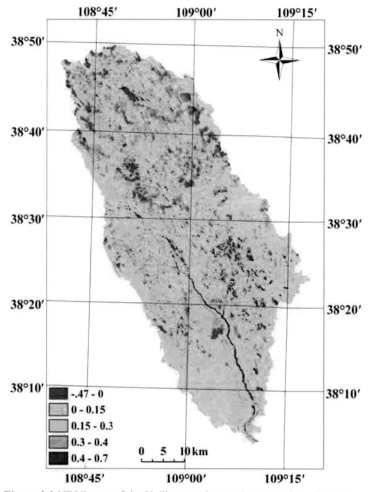

Figure 2.2 NDVI map of the Hailiutu catchment, interpretation of remote sensing data from TM image, observed on Aug, 2008; see Table 2.1 for the conversion of NDVI values to land cover classes.

Geographically, the Hailiutu catchment is a part of the Maowusu semi-desert.

However, the catchment is mainly covered by xeric shrubland (Figure. 2.2), which occupies around 88% of the surface area (Table 2.2). The crop land mixed with wind-breaking trees occupies only 3% of the total surface area. Most crop lands are located in the river valley and in the Bulang sub-catchment. Grassland areas can be found in local depressions where groundwater is near to the surface. The catchment is characterized by a semi-arid continental climate. The long-term annual average of daily mean temperature from 1961 to 2006 is 8.1°C with the highest daily mean temperature of 38.6°C recorded in 1935 and the lowest value of -32.7°C observed in 1954. The monthly mean daily air temperature is below zero in the winter time from November until March (Figure. 2.3a). The growing season starts in April and lasts until October. The mean value of the annual sunshine hours is 2926 hours (Xu, et al., 2009). The mean annual precipitation for the period 1985 to 2008 is 340 mm/year, the maximum annual precipitation at Wushenqi is 616.3 mm/year in 2002, and the minimum annual precipitation is 164.3 mm/year in 1999 (Wushenqi meteorological station monitoring data, 1985–2008). Majority of precipitation occurs in June, July, August and September (Figure. 2.3b). The mean annual pan evaporation (recorded from evaporation pan with a diameter of 20 cm) is 2184 mm/year (Wushenqi metrological station, 1985–2004). The monthly pan evaporation significantly increases from April, reaches highest in May to July, and decreases from August (Figure. 2.3c). The mean monthly discharges at Hanjiamao station vary from 0.86 m^3/s in April to 11.6 m^3/s in August (Figure. 2.3d).

Table 2.2 Land cover in Hailiutu catchment.

Land cover	NDVI	Area (km^2)	Percent
Bare soil or constructed area	≤ 0	148	5.6
Low density shrubland	$0 < NDVI \leq 0.15$	1656	62.6
High density shrubland	$0.15 < NDVI \leq 0.3$	669	25.3
Grassland	$0.3 < NDVI \leq 0.4$	90	3.4
Crop land and trees	$0.4 < NDVI \leq 0.7$	82	3.1
Total		2645	100

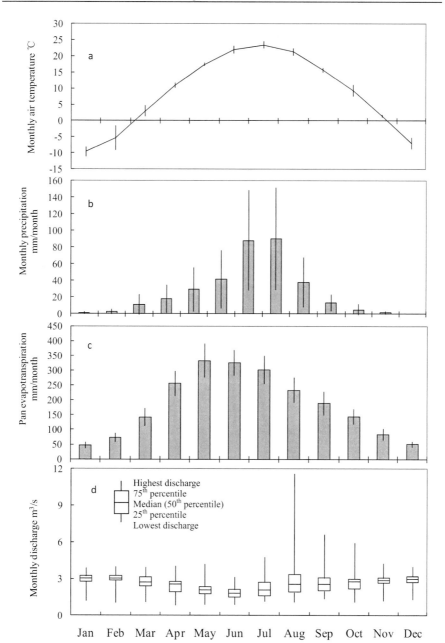

Figure 2.3 (a) mean monthly air temperature (2005–2008); (b) Mean monthly precipitation (1984–2005); (c) mean monthly pan evaporation (1984–2005) at Wushenqi meteorological station; and (d) mean monthly discharge at Hanjiamao station (1957–2007), the error bars indicate the standard deviations for precipitation, potential evaporation, air temperature and the percentiles of the discharge.

2.2.2 Data

There are four meteorological stations situated in and around the Hailiutu catchment (Figure. 2.1), and one hydrological station with daily discharge measurements from 1957 to 2007 at the outlet of the catchment. Daily precipitation at meteorological stations, air temperature from 1961 to 2006 at Yulin and Hengshan, monthly pan evaporation from 1978 to 2004 at Yulin, and monthly pan evaporation from 1985 to 2008 at Wushenqi were collected for analyses (Table 2.3). Four characteristic time series were derived from daily river discharge series for flow regime shift detection. Annual mean discharge is the average of daily discharges in each year, which is used to analyze changes in mean flows. Annual maximum discharge is the maximum daily discharge in every year, which is used to analyze the changes in peak flows. Annual mean monthly minimum discharge is the average of minimum daily flow of 12 months in every year, and is used to analyze changes in low flows. Annual mean monthly standard deviation is the average of standard deviations of daily discharges in every month, and is used to analyze the variation in monthly mean flows. The climatic variables such as annual precipitation, annual total heavy precipitation (daily rainfall >10 mm/d), number of days of heavy rainfall, average air temperature in the growing season from April to October, and annual pan evaporation were analyzed for possible climate changes. The crop area data from 1949 to 2007 was collected in the Yuyang district, a part of which is located in the Hailiutu catchment. The crop area is an indicator of land use change and the related water use for irrigation. The amount of river water diversion and groundwater extraction rates was not measured and recorded; thus were not available for analysis. `

Table 2.3 List of available data in the research catchment.

Type	Name	Elevation m (a.m.s.l)	Data information
Hydrological station	Hanjiamao	1037	daily discharge from 1957 to 2007; daily river stage (1958~1960; 1963~1966; 1979~1981; 1991~1997; 2000)
Meteorological stations	Hanjiamao	1037	daily rainfall in 1961-1963; 1970-2007
	Henan	1247	monthly rainfall and pan evaporation 1985-2004
	Wushenqi	1302	monthly rainfall and pan evaporation 1985-2008
	Hengshan	1036	daily rainfall, relative humid, wind speed, air temperature 1961-2006
	Yulin	1082	daily rainfall, relative humid, wind speed, air temperature 1961-2006

2.2.3 Methods of regime shift detection

Among all the methods of detection of the regime shifts or the change points, detection of shifts in the means is the most common type (Rodionov, 2005). In this paper, the Pettitt test (Pettitt, 1979), cumulative sum chart (CUSUM) with bootstrap analysis(Taylor, 2000, Taylor, 2000), regime shift index (RSI) calculated by a sequential algorithm of the partial CUSUM method combined with the t-test (Rodionov, 2004) were applied for detecting the shifts in the means of hydrological and climatic variables and crop areas.

The Pettitt test is a non-parametric trend test for identification of a single change point in the time series data, which is often used to detect abrupt changes in hydrological series (Love, et al., 2010). The tests were carried out by the software Datascreen (Dahmen and Hall, 1990) at a probability threshold of $p = 0.8$. The time series data were manually divided into sub-series in order to detect the change points in different periods. Abu-Taleb et al. (2007) used the CUSUM and bootstrap analysis for examining annual and seasonal relative humidity variations in Jordan, which begins with the construction of the CUSUM chart for the data sets, then calculates the confidence level by performing a bootstrap analysis for the apparent changes. A sudden change in the direction of the CUSUM indicates a sudden shift in the average. Rodionov (2004) proposed a sequential algorithm that allows for early detection of a regime shift and subsequent monitoring of changes in its magnitude over time. Start with initial subdivision of the data at certain point with the predefined cut-off length, the RSI method estimates the regime shift by statistically testing the means of the previous subsets and subsequent data sets. Then continuously increases the number of the subsequent data sets and recalculates the means until the difference in the means is statistically significant. The regime shift point is identified when the means are statistically different. This point is considered as a possible starting point of the new regime. Among the methods introduced above, the Pettitt test is widely used for examining the occurrence of a single change point in the time series, while multiple change points can be detected by CUSUM and RSI methods. In this study, these three methods were used to detect the regime shifts. A change point is accepted only, if at least two methods detected the same point. Furthermore, in order to verify this change point, the Student t test in difference in two means was used to ascertain that the mean in the sub-set before the change point and the mean in the sub-set after the change point are statistically significantly different. The periodic characteristics of hydrological variables can be analyzed by harmonic series (Zhou, 1996). Harmonic analyses for monthly discharges and standard deviations of the discharge were applied for different periods based on the preliminary analyses of the flow regime shift. The periodicity, magnitude and phase shift of the harmonic components were investigated in order to detect shifts in the periodic characteristics. Finally, flow duration curves (FDC) of different periods were constructed for investigating the changes of daily discharge characteristics.

2.3 Results and discussion

2.3.1 Detection of regime shifts

2.3.1.1 River flow regime changes

The change points for flow regime shifts in 4 characteristic river flow series were detected and are listed in Table 2.4. Figure 2.4 presents the detection results of flow regime shifts with the characteristic series (solid line curve) and their step trends (dashed line). Four change points were detected in the annual mean discharge, these changes occur in 1968, 1986, 1992 and 2001. Three change points were detected in the annual maximum discharge: 1971, 1988 and 2000. The annual mean monthly minimum discharge had four change points in 1965, 1985, 1991 and 1998. Finally, the annual mean monthly standard deviation had three change points in 1971, 1990 and 2000. The change points of the annual mean monthly standard deviation are corresponding well to the annual maximum discharge, while the change points in the annual mean discharge are caused not only by the maximum and minimum flows, but also the average flow.

Table 2.4 Results of flow regime shift detection.

Time series	Timing of the change points					
Annual mean	1967 1968	1986 1988 1992	2001			
1957 ———————— ▽ ▽ ———————————— ▽ ▽ ▽ ———————— ▽ ——————— 2007						
	Pettitt	Pettitt	Pettitt	Pettitt		
	CUSUM	CUSUM	CUSUM	CUSUM		
	RSI	RSI	RSI			
Annual maximum	1971	1988	2000			
1957 ———————— ▽ ————————————————— ▽ ———————— ▽ ——————— 2007						
	Pettitt	Pettitt	Pettitt			
	CUSUM	CUSUM	CUSUM			
	RSI					
Annual mean monthly minimum	1960 1965 1967	1985	1991	1998		
1957 —— ▽ —— ▽ ▽ ——————— ▽ ———— ▽ ——— ▽ —— 2007						
	Pettitt	Pettitt	Pettitt	Pettitt		
	CUSUM	CUSUM	CUSUM	CUSUM		
RSI	RSI		RSI	RSI		
Annual mean monthly standard deviation	1971	1990	2000			
1957 ———————— ▽ ————————————————— ▽ ———————— ▽ ——————— 2007						
	Pettitt	Pettitt	Pettitt			
	CUSUM	CUSUM	CUSUM			
	RSI	RSI	RSI			

The black line represents the time from 1957 to 2007, the triangle symbols above the time line stand for change points detected at the year, the detection methods for the change points are summarized below the time line.

The harmonic analysis of monthly mean discharge time series shows the distinct periodic characteristics in the identified 5 periods (Table 2.5). Two discharge peaks occur in every year in the first period from 1957 to 1967: a smaller peak in February-March and a larger peak in August-September (Figure 2.5). The small peak in the winter period reflects the maximum groundwater discharge, while the large summer peak resulted from direct rainfall-runoff conversion. In the second period from 1968 to 1985, the winter peak is shifted one month earlier to January-February, while the summer peak is delayed by one month to October (Table 2.5). The magnitude of the summer peak became smaller than the winter peak, which is a clear indication of effects of the reservoirs and water diversions on the monthly flow regime. The summer peak disappears in the third period from 1986 to 1991, resulting in only annual periodic fluctuation with the peak flow occurring in November-January produced by groundwater discharge. The monthly flow regime in the fourth period from 1992 to 2000 resembles the third period. However, the amplitude of the variation is much smaller, which indicates increased irrigation water consumption by water diversions from the river and groundwater extraction. The amplitude of the annual periodic fluctuation is increased due to the increase of groundwater discharge in November to January in the fifth period from 2001 to 2007.

Table 2.5 Harmonic characteristics of monthly discharge and standard deviation at Hanjiamao station.

Monthly mean		1957-1967	1968-1985	1986-1991	1992-2000	2001-2007
	Periodicity (month)	6	12 / 6	12	12	12
	Amplitude	0.52	0.47 / 0.26	0.58	0.35	0.71
	Peak discharge	Feb-Mar	Jan-Feb/	Dec-Jan	Dec-Jan	Dec-Jan
	occurs in	Aug-Sep	Oct			
Standard deviation		1957-1970	1971-1989		1990-1999	2000-2007
	Periodicity (month)	12 / 6	12		12	12 / 6
	Amplitude	0.80 / 0.44	0.39		0.15	0.46 / 0.25
	Peak discharge	Feb	Jul		Jul	Feb
	occurs in	Aug				Aug

Three change points were detected in mean monthly standard deviation series, which, therefore, was divided into 4 periods. There are two peaks in mean monthly standard deviation of discharge time series in the first period from 1957 to 1970, one peak occur in February and another in August (Figure 2.5). The amplitude of the peak in August is very large indicating large discharge variations during the summer rainy season. Only one 12-month periodicity with an amplitude of 0.39 m³/s and peak in July was found in second period from 1971 to 1989. The similar harmonic characteristics were found in the third period from 1990 to 1999 except for a smaller amplitude of 0.15 m³/s. The variations of river discharges were substantially reduced in these two periods. The fourth period from 2000 to 2007 shows the similar periodic changes with two harmonics as in the first period; however, the summer peak (in August) is smaller (Figure 2.5).

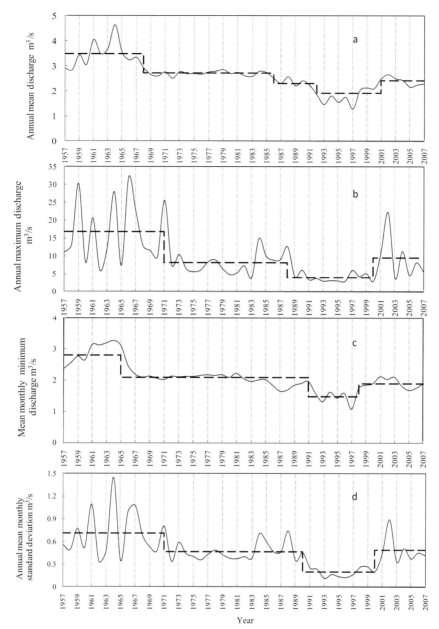

Figure 2.4 Flow regime shifts in the annual mean discharge (a), annual maximum discharge (b), annual mean monthly minimum discharge (c), and the annual mean monthly standard deviation (d) at Hanjiamao station from 1957 to 2007, the solid lines are the characteristic series and the dashed lines are the their step trends.

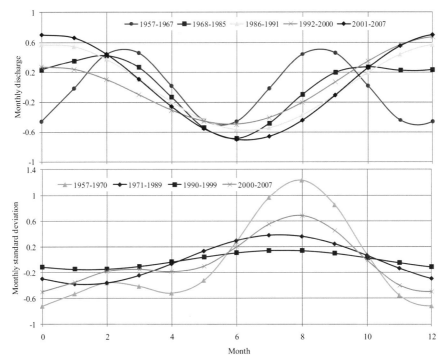

Figure 2.5 Harmonic changes in monthly mean discharge and standard deviation at Hanjiamao station for different periods.

2.3.1.2 Past climate and land use change

The correlation coefficients among monthly precipitation at four meteorological stations are more than 0.83 (Table 2.6). The correlation coefficients between annual pan evaporation at Yulin and Wushenqi are also higher than 0.78. The correlation coefficient between the annual average air temperature at Yulin and Hengshan is 0.98. The high correlation coefficients indicate that the annual precipitation, pan evaporation, and air temperature at the meteorological stations are spatially homogeneous and consistent.

Table 2.6 Correlation coefficients of monthly precipitation at meteorological stations.

Correlation coefficients	P_{Yulin}	$P_{Wushenqi}$	P_{Henan}	$P_{Hengshan}$
P_{Yulin}	1			
$P_{Wushenqi}$	0.89	1		
P_{Henan}	0.88	0.84	1	
$P_{Hengshan}$	0.89	0.83	0.95	1

The results of regime shift detection of climate variables and crop area in Hailiutu catchment are shown in Table 2.7 and Figure 2.7. No significant shifts were found in the annual precipitation, total heavy precipitation (daily rainfall >10mm/d), number of days of the heavy precipitation, and pan evaporation time series by all three methods. Since the temperature from November until March is below zero, ground freezes in winter season. There are also no ever-green plants in the catchment. Thus evaporation can be assumed zero in winter season. The growing season starts from April and ends in October in this region. The temperature in the growing season has impact on actual evaporation. Therefore, average temperature in growing season (from Aril to October) was used for the detection of temperature changes. A significant increase in the average temperature in the growing season was detected at 1997 at Yulin and Hengshan meteorological stations. Three change points of crop area were detected at 1971, 1990, and 1999. The crop area was decreased by 7% from the period of 1957-1970 to the period of 1971-1989. The crop area was increased by 15% from the period of 1971-1989 to the period of 1990-1998. The crop area in the last period of 1999-2007 was decreased by 17% from the previous period.

Table 2.7 Regime shift detection results of climate variables and crop area.

Time series	Timing of the change points		
Average temperature from April to October at Yulin			
	1985	1997	
	▽	▽	
1957 ——————————	Pettitt	Pettitt	———— 2007
		CUSUM	
		RSI	
Average temperature from April to October at Hengshan			
	1985	1997	
	▽	▽	
1957 ——————————	Pettitt	Pettitt	———— 2007
		CUSUM	
		RSI	
Crop area in Yuyang			
	1970 1971	1990 1991	1999
	▽ ▽	▽ ▽	▽
1957 ——————————	Pettitt	Pettitt	Pettitt ———— 2007
	CUSUM	CUSUM	CUSUM
	RSI	RSI	RSI

The black line represents the time from 1957 to 2007, the triangle symbols above the time line stand for change points detected at the year, the detection methods for the change points are summarized below the time line.

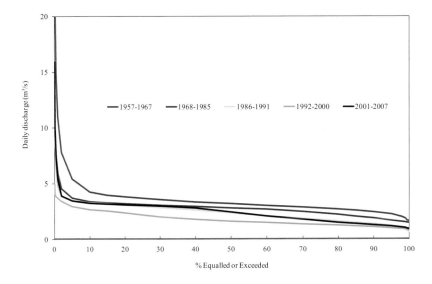

Figure 2.6 Flow duration curves for mean daily discharges at Hanjiamao station in the 5 different periods.

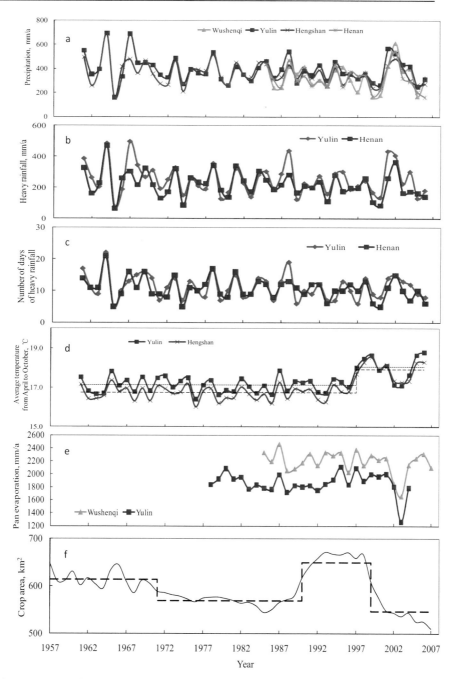

Figure 2.7 Annual precipitation (a), Heavy precipitation (>10mm/d) (b), number of days of heavy precipitation (c), annual mean temperature from April to October (d) at meteorological stations from 1961 to 2006. Annual pan evaporation (e) and annual crop area (f) in Yuyang district from 1957 to 2007, the dashed lines are the step trends.

2.3.2 Analysis of the results

2.3.2.1 Characteristics of flow regime changes over time

According to the change point detection results, the annual mean flow regime at Hanjiamao station from 1957 to 2007 can be divided into 5 distinctive periods. The flow regime in the first period from 1957 to 1967 represents quasi natural variations with a mean discharge value of 3.49 m³/s, a 6 month periodicity, and the comparable high daily discharges. Two peaks in monthly mean discharge series reflect a small discharge peak in winter with maximum baseflow from groundwater discharge and a large discharge peak in summer originating from seasonal rainfall. In the second period from 1968 to 1985, the flow regime can be characterized by a lower mean discharge value of 2.72 m³/s, 12 and 6 month periodicities, and lower daily flows, which implies that the discharge had been affected. In the third period of 1986-1991, the mean discharge further decreased to 2.3 m³/s. The magnitude of the summer peak discharge became smaller than the winter peak. The daily flows continuously declined in the fourth period of 1992–2000. The discharge reached lowest levels with a mean value of 1.92 m³/s, lowest amplitude of harmonic components, and lowest daily flows. The discharge recovered comparably to the previous but not yet to the natural level in the fifth period of 2001–2007 with a mean value of 2.41 m³/s, 12 month periodicity, and comparable higher daily flows.

2.3.2.2 Causes of flow regime shifts

The climatic variables such as precipitation, pan evaporation, and the human activities are normally correlated with the observed discharge variability. Hu et al. (2012) analyzed trends in temperature and rainfall extremes from 16 meteorological stations in the Yellow River source region. They found significant warming trends in the whole source area, but no significant changes in the annual rainfall for the majority of the stations except in the upper part of the region. Thus, the decrease of the stream flow in the source region of the Yellow River can be ascribed to the increase of temperature resulting in higher evaporation losses. In this study, no significant changes in the annual precipitation, the heavy rainfall (daily rainfall >10 mm/d), and the number of heavy rainfall days, were detected. There are no significant changes detected in pan evaporation rates. The change point detection methods indentified an increase of average growing-season temperature since 1997. This temperature increase could also have an impact on the decrease of river flows since the actual evaporation was expected to increase with a higher temperature. On the contrary, river flow has increased since 2001. The main cause was the significant decrease of crop area with the implementation of the policy to return farmland to forest and grassland that started in 1999. In the study catchment, crop area in large parts returned to the natural vegetation cover, which consists of desert bushes. Actual evaporation from desert bushes is much smaller than from irrigated cropland under a higher temperature. The decrease of net water use because of large decrease in crop area had much larger impact on the river flow than a possible increase of actual evaporation from desert bushes under higher temperature. This explains why river flows have increased in spite of the temperature increase. Therefore, the flow regime shifts in the Hailiutu River were not likely caused by climatic changes.

Figure 2.8 connects the local/regional land policies with the changes in crop area and observed flow regime shifts in Hailiutu River. The changes of crop area in Yuyang district can represent the changes in the amount of water diverted from river and groundwater exploitation for irrigation. The crop area has been influenced by the policy for cultivation and agriculture during the last 51 years (Figure 2.8). The crop area remained at a certain level from 1957 to 1970 when the crop land development and degradation processes were balanced. The crop area gradually decreased from early 1970's when the cultivated land was converted to terraces for agriculture with the policy of emulating Dazhai on agriculture campaign (Bi and Zheng, 2000). The policy of distributing the cultivated land to farmers was implemented in the early 1980's, which eventually stimulated the farmers to enlarge the crop area (Zhang, et al., 2009). The crop area was increased during 1990-1998 for the vegetable production according to the policy of non-staple food supply in urban districts in 1988 (Wang and Xu, 2002), which implies that intensive water diversion works and groundwater extraction wells were constructed to secure the irrigation for the crop land. The Chinese government has implemented a program to return the crop land to forest or grassland for ecosystem rehabilitation since 1999 (Li and Lv, 2004), which resulted in a decrease of crop area from 2000 to 2007. Land cover change caused by the cultivation policy is related to crop area change in the study area. In the river valley, it is the change between maize and nursery garden for trees. Trees (especially young, fast growing trees) may consume more water, but the area is limited. In the upper part of the catchment, it is the change between maize and desert bushes. Desert bushes (not irrigated) consume much less water compared to irrigated maize. We can

29

conclude that with a large decrease of crop area since 1999 (Figure 8), actual evaporation was decreased resulting in an increase of river discharge. Although temperature increased in the same period, but the increase of actual evaporation by the temperature increase was much less than the reduction of actual evaporation by the crop area decrease.

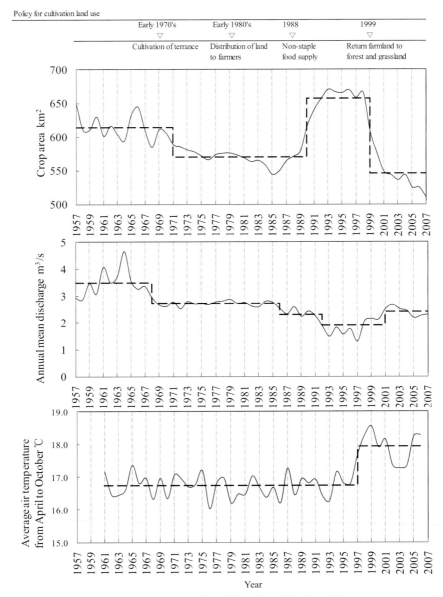

Figure 2.8 Illustration of the causes for regime shifts; the cultivation policy, crop area, annual discharge, and temperature form April to October at Yulin meteorological station.

The hydraulic works such as reservoirs may significantly retard and reduce surface runoff and, hence, are expected to have great impact on the discharge variability and particularly on the annual maximum streamflow (Li, et al., 2007). The change point in the annual maximum discharge was detected at 1971 when the two reservoirs were constructed in the main river and tributary. In the subsequent period of 1971–1984, the standard deviation decreased due to the combined effects from the reservoirs and groundwater exploitation. The maximum discharge and standard deviation were slightly increased in the period of 1985–1990 when less operation of the hydraulic engineering works during the initial stage of policy that the farmers could cultivate individually (Zhang, et al., 2009). In the fourth period of 1990–1999, the maximum discharges are lowest due to the combined effects of reservoirs and construction of several water diversions for irrigation. The increase of maximum discharge and the standard deviation in the fifth period indicate a reduction of water diversion.

The low flows during the dry season are sensitive to changes in the groundwater system because the groundwater discharge dominates in the dry season. The groundwater abstractions started in the 1970's based on the analysis of annual mean monthly minimum discharge. The intensive groundwater exploitation is responsible for the lowest minimum discharge in the period of 1992–2000. The increase of the minimum discharge in the fifth period indicates a possible reduction of groundwater extraction.

2.3.2.3 Regression analysis

The lack of historical data on river water diversions, groundwater abstractions, detailed land cover changes and actual evaporation prevent a full quantitative cause-effect analysis. Nevertheless, the simple regression analysis of annual mean discharge against annual precipitation, crop area and annual average air temperature in growing season was performed to get insight on their cause-effect relations. Based on the results of detection for flow regime shifts, the flow regime has been disturbed since 1968. The annual mean discharge at the Hanjiamao station, annual precipitation and air temperature in growing season at the Yulin station, and crop area in Yuyang district from 1968 to 2006 were selected for the analysis. Table 2.8 shows the correlation coefficients among the variables. It is clear that the annual mean discharge is positively dependent of the precipitation, but negatively dependent of crop area and air temperature. It can further be seen that the correlation coefficient between river discharge and crop area is much larger than those of precipitation or temperature.

The regression equation among the discharge, precipitation, air temperature in growing season, and crop area was found as follows:

$$Q = 124.8 + 0.004 \times P - 2.904 \times T - 0.079 \times A_{crop}$$
(2.1)

Table 2.8. Correlation coefficients among discharge, precipitation, air temperature in growing season, and crop area from 1968 to 2006.

	Q (mm/year)	P (mm/year)	A_{crop} (km^2)	T (°C)
Q (mm/year)	1.00			
P (mm/year)	0.42	1.00		
A_{crop} (km^2)	-0.71	-0.40	1.00	
T (°C)	-0.33	-0.21	-0.05	1.00

Q is the annual mean discharge at Hanjiamao station, P and T are the annual precipitation and temperature in growing season at Yulin station, A_{crop} is crop area in Yuyang district.

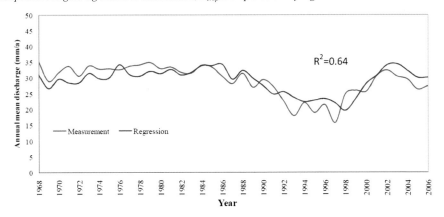

Figure 2.9 Fit of prediction by multiple regression of the annual mean discharge at Hanjiamao station with the climatic variables at Yulin station and crop area in Yuyang district from 1968 to 2006.

Where the Q is the annual mean discharge in mm/year, P is annual precipitation in mm/year, T is annual average temperature in °C, and Acrop is the crop area in km^2. The multiple regression analysis (Figure 2.9) yielded a coefficient of determination of 0.64, indicating that 64% of the variation in the annual mean discharge can be explained by the combined variation of the precipitation, crop area and temperature in growing season. With an increase of crop area or temperature, river discharge decreases since the actual evaporation increases with the increased temperature or crop area. The regression could be likely improved if the actual data of water diversion and groundwater extraction would be available.

2.4 Conclusion

The flow regime shift detection and harmonic analysis show that the flow regime of the Hailiutu River has been changed dramatically over the last 51 year. Four major shifts in the flow regime have been detected in 1968, 1986, 1992 and 2001. The first period from 1957 to 1967 represents in general the natural variation of the river flow with higher annual mean discharge, large annual maximum discharge, large standard deviation of mean monthly discharges, 6-month periodicity of two flow peaks (one in winter and one in summer) per year, and the high daily flow variability. The flow regime is modified in the second period from 1968 to 1985 mainly by the construction of reservoirs and water diversion works, which resulted in lower annual mean discharge, significant reduction of summer flow peak, smaller standard deviation of mean monthly discharges, and low daily flows. In the third period of 1986–1991, the discharge continuously decreased due to the combined impacts by the river water diversion and the increase of the groundwater extraction. The summer flow peak is vanished because of river water diversion, leaving only the winter flow peak formed by groundwater discharge. The winter flow peak is shifted one month earlier possibly due to the irrigation return flow in the river valley, which still needs to be investigated. The annual mean discharge, annual maximum discharge, and the annual average standard deviation of mean monthly discharge were lowest in the fourth period of 1992–2000 due to a large increase of crop land with intensive water diversion and groundwater extraction for irrigation. The flow rate and variation recovered, but not yet to natural levels in the fifth period of 2001–2007, which can be attributed to the decrease of water extraction as a result of the implementation of the policy of returning farmland to forest and grassland in the catchment. In the study catchment, the policy of return farmland to forest and grassland means to convert cropland to natural xeric bushland. The bushland is not irrigated and evaporates less than irrigated cropland, which seems to be the main reason for the streamflow increases from 2001 onwards.

The analyses show that there are no significant climate changes in the Hailiutu River catchment. Only the average air temperature in the growing season (April to

October) was increased since 1997, but has not caused significant changes in the flow regime. On the contrary, historical land use policy changes had always had the footprint in the flow regime changes. Reservoirs and river water diversions are the main causes of reduction and disappearance of summer flow peaks. Groundwater extraction for irrigation reduces river base flow and contributes to the decrease of the annual mean discharge. The simple regression analysis of annual mean discharge against annual precipitation, crop area and average air temperature in growing season indicates that the largest contribution to the variation in river discharges comes from crop area, the combined effects of temperature and precipitation on flow variation is smaller than that of the crop area.

Expansion of the crop area in 1989-1999 not only caused lowest river flow, but the decrease of regional groundwater levels threaten the health of the desert vegetation. The degradation of the desert vegetation ecosystem forced the local government to implement the policy of return farmland to desert vegetation in order to reduce groundwater abstraction for irrigation since 1999. This is the only time in the study catchment that the human policy was changed in response to the hydrological regime change. The positive effects have been observed since 2001: river flow is recovered; vegetation health is improved.

The river flow regime changes might have consequences on riparian ecosystems and downstream water use. The sustainable water resources development and management in the Hailiutu river catchment must consider the interactions of groundwater and river flow, and the consequences on the vegetation and downstream water use. Future research work should analyze effects of groundwater extraction on river flows and groundwater dependent vegetation in the catchment.

3. A multi-method approach to quantify Groundwater/surface water interactions *

Abstract: Identification and quantification of groundwater and surface-water interactions provide important scientific insights for managing groundwater and surface-water conjunctively. This is especially relevant in semi-arid areas where groundwater is often the main source to feed river discharge and to maintain groundwater dependent ecosystems. Multiple field measurements were taken in the semi-arid Bulang sub-catchment, part of the Hailiutu River basin in Northwest China, to identify and quantify groundwater and surface-water interactions. Measurements of groundwater levels and stream stages for a 1-year investigation period indicate continuous groundwater discharge to the river. Temperature measurements of stream water, streambed deposits at different depths, and groundwater confirm the upward flow of groundwater to the stream during all seasons. Results of a tracer-based hydrograph separation exercise reveal that, even during heavy rainfall events, groundwater contributes much more to the increased stream discharge than direct surface runoff. Spatially distributed groundwater seepage along the stream was estimated using mass balance equations with electrical conductivity measurements during a constant salt injection experiment. Calculated groundwater seepage rates showed surprisingly large spatial variations for a relatively homogeneous sandy aquifer.

3.1 Introduction

Groundwater and surface water have been managed as isolated components for a long time, but they are hydrologically connected in terms of both quantity and quality (Winter, 1999). Thus, a better understanding of the interactions between groundwater and surface water could provide crucial scientific insights for integrated management of water resources. Interactions between groundwater and surface water such as in

*This chapter is based on paper A multi-method approach to quantify groundwater/surface water-interactions in the semi-arid Hailiutu River basin, northwest China. Yang Z, Zhou Y, Wenninger J, Uhlenbrook S (2014). Hydrogeology Journal 22: 527-541. DOI 10.1007/s10040-013-1091-z.

headwaters, streams, lakes, wetlands, and estuaries have been studied since the 1960s (Winter, 1995, Woessner, 2000, Sophocleous, 2002). Many methods of quantifying the interactions between groundwater and surface water have been applied by researchers all over the world. The measuring methods for groundwater and surface-water interactions were summarized by Kalbus et al.(2006), Brodie et al.(2007), Rosenberry and LaBaugh(2008). A number of studies have been conducted for identifying the interactions between groundwater and surface water using: multiple field investigations (Oxtobee and Novakowski, 2002, Langhoff, et al., 2006, Rautio and Korkka-Niemi, 2011); differences in hydraulic heads between river and groundwater (Anderson, et al., 2005) with supplemental chemical and stable isotope data (Oxtobee and Novakowski, 2002, Marimuthu, et al., 2005); temperature studies of streambeds (Conant, 2004, Schmidt, et al., 2007, Westhoff, et al., 2007, Vogt, et al., 2010, Lewandowski, et al., 2011); and chemical and isotopic tracers (Ayenew, et al., 2008, Didszun and Uhlenbrook, 2008, Wenninger, et al., 2008). Field surveys such as point measurements of hydraulic heads are normally used for interpreting the interaction between groundwater and surface water. Seepage meters can directly measure the exchange between the groundwater and surface water (Landon, et al., 2001). The stable isotopes deuterium and oxygen-18 as well as hydrochemical tracers have been widely used for investigating runoff generation processes (Uhlenbrook and Hoeg, 2003), determining the hydrological exchange (Ruehl, et al., 2006), and inferring groundwater/surface-water exchanges (Rodgers, et al., 2004). There are a number of publications dealing with the application of tracers such as hydrochemical components (Wels, et al., 1991, Uhlenbrook, et al., 2002, Kirchner, 2003) or environmental isotopes (Sklash and Farvolden, 1979, McDonnell, et al., 1990) in hydrological investigations. Most of them have been employed for separating the stream flow into pre-event and event water using environmental isotopes or spatially distinct components of surface water and groundwater using hydrochemical compounds (Uhlenbrook and Hoeg, 2003). As pointed out by Anderson (2005) and Constantz (2008), heat can be used as a tracer for investigating the groundwater and surface-water connectivity. The temperatures beneath the streambed have been interpreted for delineating and quantifying groundwater discharge zones, estimating the vertical velocity and fluxes in the streambed. Constantz (2008) distinguished the thermal pattern beneath the streambed for losing stream reaches and gaining reaches. The oscillating surface temperature signal results in both conductive and advective

36

heat transfer, with higher infiltration rates associated with a losing stream reach resulting in greater advection, deeper penetration, and shorter lags in temperature extremes at a given depth. However, using streambed temperatures to quantify groundwater/surface-water interactions is limited to locations and time periods where groundwater and stream water have sufficient temperature differences (Schmidt, et al., 2006). The interaction between groundwater and surface water can be quantified by directly measuring the differences of discharge along the reach with e.g. current meters (Becker, et al., 2004). Discharge measurements using a continuous injection of a sodium chloride (NaCl) solution and integration of the electrical conductivity as a function of time is a traditional and well-documented method for turbulent streams (Hongve, 1987). The NaCl dilution and mass recovery method were employed for estimating the channel water balance (Payn, et al., 2009) as well as net change in discharge, gross hydrologic loss, and gross hydrologic gain in experimental reaches by assuming constant discharge and complete mass recovery.

In China, several studies have been conducted for quantifying interactions between groundwater and surface water in arid and semi-arid regions with tracer methods (Wang, et al., 2001, Nie, et al., 2005, Wang, et al., 2009), water balance (Xiao, 2006), and model simulations (Wang, et al., 2006, Hu, et al., 2007, Hu, et al., 2009). An integrated approach that consists of multiple measuring methods has not yet been applied to investigate groundwater and surface-water interactions.

Water level and temperature measurements indicate that groundwater discharges to the river or groundwater can be recharged by the river. Discharge measurements and hydrograph separation can estimate the total groundwater discharge in the sub-catchment. However, determining spatial variations of groundwater discharges along rivers is still a challenge. The application of distributed temperature sensors along river reaches identified large spatial variations (Selker, et al., 2006, Lowry, et al., 2007); however, the method cannot provide estimations of groundwater discharges directly. Seepage meters were used to measure groundwater fluxes at multiple points (Paulsen, et al., 2001, Rosenberry and Morin, 2004, Rosenberry and Pitlick, 2009, Hatch, et al., 2010), but cannot estimate total groundwater discharge of the total stream reach. Temperature gradient measurements beneath the streambed combined with the analytical (Silliman, et al., 1995, Becker, et al., 2004) or numerical heat transport model (Constantz, 1998, Constantz and Stonestrom, 2003), can only estimate point groundwater discharges.

This study used multiple field measurements to identify and quantify groundwater and surface-water interactions in a small catchment located in the semi-arid Erdos Plateau in Northwest China. Measurements of groundwater and surface-water levels could provide an indication of the directions of water exchange between groundwater and surface water. Measurements of temperature in groundwater, stream water, and streambed deposits at different depth give direct evidence of connectivity and flow directions between groundwater and stream water at specific locations. A hydrograph separation analysis of stream discharges using stable isotopes as a tracer estimates relative contributions of groundwater and direct surface runoff to the total stream discharges for a rainfall event. A new method to quantify spatial distribution of groundwater seepage in a gaining stream using electrical conductivity profile measurements along the stream course before and during a constant salt injection is developed and tested. The successive solutions of mass balance equations along the measured stream segments provide estimates of groundwater seepage for every 10-m reach. This method is cost-effective and can be widely applied to quantify the groundwater seepage to streams. The spatially distributed groundwater seepage provides also new sources of data to calibrate coupled groundwater/surface-water models.

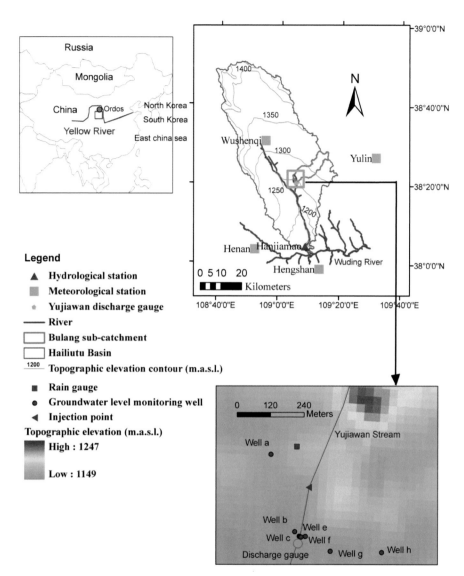

Figure 3.1 Location of the Yujiawan discharge gauging station, groundwater monitoring wells, rain gauge, and the constant injection point at the Yujiawan stream in the Bulang sub-catchment inside the Hailiutu River basin.

3.2 Material and methods

3.2.1 Study site

The Bulang catchment is a sub-catchment of the Hailiutu River basin in the middle section of the Yellow River basin in Northwest China (Figure 3.1). The total drainage area of the Bulang sub-catchment is 91.7 km². The surface elevation of the Bulang sub-catchment ranges from 1,300 m at the northeastern boundary to 1,160 m above mean sea level at the catchment outlet in the southwest. The land surface is characterized by undulating sand dunes and a perennial river in the downstream area (Yang, et al., 2012). The long-term annual average of daily mean temperature is 8.1°C and the monthly mean daily air temperature is below zero in the wintertime from November until March. The mean annual precipitation for the period 1985 to 2008 was 340 mm/year, measured at the nearby meteorological station in Wushenqi. The majority of precipitation falls in June, July, August and September. The mean annual pan evaporation (recorded with an evaporation pan with a diameter of 20 cm) is 2184 mm/year (Wushenqi metrological station, 1985–2004). The monthly pan evaporation significantly increases from April onwards, reaches highest values from May to July, and decreases from August onwards. The geological formations in the Bulang sub-catchment mainly consist of four strata (Hou, et al., 2008): (1) the Holocene Maowusu sand dunes with a thickness from 0 to 30 m; (2) the upper Pleistocene Shalawusu sandstone with a thickness of 5 to ~90 m; (3) the Cretaceous Luohe sandstone with a thickness of 180 to ~330 m; and (4) the bedrock, which consists of Jurassic impermeable sedimentary formations. No continuous semi-permeable formations exist, so that the Quaternary and Cretaceous formations form a continuous regional aquifer system in the Bulang sub-catchment. The natural vegetation in the sub-catchment is dominated by salix bushes (Salix Psammophila), while the livelihood of the local people depends on growing maize on croplands. The low density shrub, high density shrub, and cropland in the Bulang sub-catchment occupy 69, 23, and 8% of the total area, respectively. Groundwater is abstracted for irrigation in the growing season from April to October in the Bulang sub-catchment. The total length of the investigated segment of the Yujiawan stream is 900 m, while the width of the stream varies from 0.5 m at the upstream section to 2.5 m at the downstream section. The depth of the water in the stream is smaller than 5 cm except during flood events. The stream bank and the streambed are formed by sand.

3.2.2 Groundwater and stream stage monitoring

A set of groundwater-level monitoring wells was installed at the streambed, stream banks, flood plain, and the terrace at the upstream of the discharge gauge in the

Bulang sub-catchment (Figure 3.1 and 3.2). Groundwater levels in the shallow aquifer were measured in 10 minute intervals using a MiniDiver submersible pressure transducer (Eijkelkamp Agrisearch Equipment, Giesbeek, The Netherlands) installed in the monitoring wells. Barometric compensation was carried out using air pressure measurements from a BaroDiver (Eijkelkamp Agrisearch Equipment, Giesbeek, The Netherlands) installed at the site. Groundwater levels were reported as the height of the water table above mean sea level with the calibration of the land elevation of wells, height of water column above the MiniDiver in the wells, and the depth of the MiniDiver in the boreholes. The rainfall was recorded by the a rain gauge (HOBO RG3, Onset Corporation, Bourne, USA) near the discharge gauging station (Figure 3.1).

3.2.3 Temperature

Streambed temperatures at depths of 10, 30, 50 and 80 cm were measured by HOBO Pro v2 water temperature sensors (U22-001, Onset Corporation, Bourne, USA) installed in a 5-cm diameter polyvinyl chloride (PVC) pipe (Figure 3.2) upstream of the discharge gauge. In order to avoid vertical fluxes in the pipe, the two ends of the PVC pipe were sealed and several PVC blockers were put in the pipe between the sensors. The temperature sensors can register the temperature of the surrounding streambed through horizontal holes at different depths. Water temperature at the surface of the stream bottom is recorded by a temperature sensor on the surface of the streambed. One piezometer, equipped with a MiniDiver, was installed in the streambed for measuring the groundwater level below the stream. The stream stage was registered by a MiniDiver in steel stilling pipe installed in the stream, where the water level inside the steel pipe is equal to the stream stage. The sampling frequency for groundwater level, stream stage, and temperature at different depth beneath the streambed was set to 10 minutes.

3.2.4 Discharge measurements

In order to measure the discharge of the Bulang sub-catchment, one discharge gauging station was constructed at the outlet of the sub-catchment (Figure 3.1). The Yujiawan gauging station consists of one permanent rectangular weir equipped with an e+ WATER L water level logger (type 11.41.54, Eijkelkamp Agrisearch Equipment, Giesbeek, The Netherlands) where water levels are recorded with a frequency of 30 minutes. Water depths are converted to discharges using a rating curve based on regular manual discharge measurements carried out with a current meter and the velocity-area method.

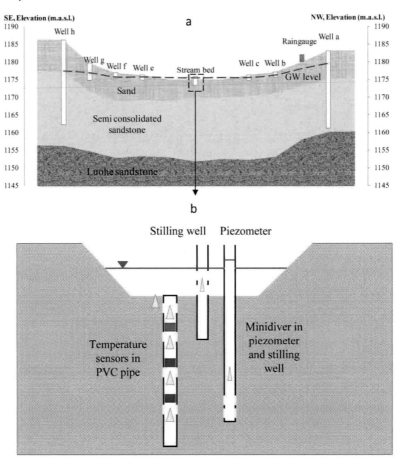

Figure 3.2 (a) Schematic plot of the groundwater monitoring wells installed in the Bulang sub-catchment; the dotted blue line indicates groundwater heads in the monitoring wells and the piezometer in the streambed at 20:00 14 June, 2011; (b) the installation of temperature sensors, piezometer in the streambed, and the stilling well for the stream stage registration.

3.2.5 Event sampling for hydrograph separation

A hydrograph is the time-series record of water level, water flow or other hydraulic properties, and can be analyzed to gain insights into the relationships between rivers and aquifers (Brodie, et al., 2007). The runoff event of 2 July 2011 in the Bulang sub-catchment was intensively sampled to carry out a hydrograph component separation. Water samples of precipitation, groundwater, and the stream water were collected from 1 July to 5 July 2011 in hourly intervals and analyzed for their hydrochemical and isotopic compositions. The analyzed isotopes are the stable water isotopes oxygen-18 (^{18}O) and deuterium (^{2}H). The chemical analysis includes the major anions and cations. The hydrograph separation using tracers is based on the steady-state mass balance equations of water and tracer fluxes.

3.2.6 Seepage calculation with electrical conductivity (EC) profile

3.2.6.1 Electrical conductivity (EC) profile measurements

A constant salt tracer injection experiment was conducted on 21 June 2011 along a 180 -m river profile upstream of the discharge gauge in the Yujiawan stream (Figure 3.3). A Mariotte bottle was used for injecting the salt tracer at a constant rate (Moore, 2004). The constant injection rate from the Mariotte bottle was manually calibrated with a stopwatch and beaker in the field. A sodium chloride (NaCl) solution was used for the constant injection experiment. The electrical conductivity was measured along the 180m profile with intervals of 10 m from the injection point to the downstream discharge gauge before and during the constant injection experiment. The electrical conductivity values of stream water were measured with a portable water quality multi-meter (18.28 and 18.21.sa temperature/conductivity meter, Eijkelkamp Agrisearch Equipment, Giesbeek, The Netherlands). The background EC profile was measured from 10:00 to 11:00 on 21 June 2011. The constant injection experiment started at 11:10, the stable plateau value of the EC at the discharge gauging station was observed around 11:51. From 11:40 to 14:00, the EC profile of the stream under the constant injection experiment was measured every 10 m along the 180 m stream reach. The width of the stream varies from about 0.5 m at the first 70 m downstream of the injection point to 2 m at the discharge weir. The discharge remained approximately constant during the experiment. The mixing length is less than the 10 m from the theoretical estimation, and was confirmed by measuring the EC value at different locations along the downstream river cross section.

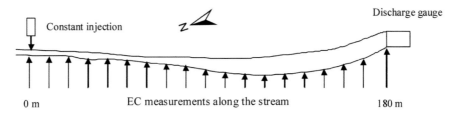

Figure 3.3 Plan view of locations of electrical conductivity (EC) measurements in the Yujiawan stream.

3.2.6.2 Estimation of seepage with the natural EC profile

Two mass balance equations were formulated with EC measurements under natural conditions. The mass balance of the total 180 m stream reach (Figure 4a) was used to calculate the upstream inflow and total seepage. The mass balance per stream segment was used to calculate the seepage every 10 m (Figure 3.4b).

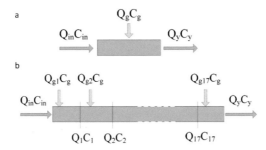

Figure 3.4 Schematic plot of mass balance calculations under the natural situation: (a) for the total 180 m reach and (b) for 10-m segments.

The mass balance equations for the total 180 m reach (Figure 4a) are:

$$Q_{in}C_{in} + Q_gC_g = Q_yC_y \tag{3.1}$$

$$Q_{in} + Q_g = Q_y \tag{3.2}$$

Where the Q_{in} is the inflow from the upstream in m^3/s; Q_y is the outflow at the Yujiawan gauging station in m^3/s; Q_g is the total groundwater seepage along the 180-m reach in m^3/s; C_{in} is the EC value of the upstream inflow water in $\mu S/cm$; C_y is the EC value at the Yujiawan gauging station in $\mu S/cm$, and C_g is the average EC value of groundwater in $\mu S/cm$. Equations (3.1) and (3.2) can be solved for Q_{in} and Q_g when the groundwater EC value is known. The average EC value of the groundwater was calculated with 7 groundwater samples collected from the observation wells near the stream. The EC values vary from 313 to 823 $\mu S/cm$ with average value of 545 $\mu S/cm$. Therefore, the Q_{in}, Q_g, and the average groundwater seepage rate along the

180-m reach can be calculated.

The mass balance equations for the first 10-m segment (Figure 3.4b) can be written as follows:

$$Q_{in}C_{in} + Q_{g1}C_g = Q_1C_1 \tag{3.3}$$

$$Q_{in} + Q_{g1} = Q_1 \tag{3.4}$$

Where the Q_{g1} and Q_1 are the groundwater seepage in the first 10-m segment and the discharge at the end of the first segment in m^3/s; C_g and C_1 are the EC values of the groundwater seepage and total discharge at the end of the first segment in $\mu S/cm$, respectively. Combining equations (3.3) and (3.4) can solve for Q_1 and Q_{g1} with the formula:

$$Q_1 = Q_{in}\frac{C_{in}-C_g}{C_1-C_g} \tag{3.5}$$

$$Q_{g1} = Q_1 - Q_{in} \tag{3.6}$$

The stream discharge and groundwater seepage at the remaining segments can be calculated successively with Equations (3.5) and (3.6).

3.2.6.3 Estimation of seepage using the EC profile under a constant injection

The EC profile measurements were taken when a constant plateau value was reached at the Yujiawan gauging station. Because of the measurement error of the extremely high EC value of the injection solution and the mixing effect, the upstream boundary is taken from 10-m downstream from the injection point. The mass balance for the downstream 170-m reach (Figure 3.5a) can be calculated as follows:

$$Q_1'C_1' + Q_g'C_g = Q_y'C_y' \tag{3.7}$$

$$Q_1' + Q_g' = Q_y' \tag{3.8}$$

Where, Q'_1 is the discharge at 10-m downstream from the constant injection point in m^3/s; Q'_y is the discharge at the Yujiawan gauging station in m^3/s; Q'_g is the total groundwater seepage along the 170-m reach in m^3/s; C'_1 is the EC value 10-m downstream in $\mu S/cm$ and C'_y is the EC value at Yujianwan gauging station in $\mu S/cm$.

The mass balance equations for the second segment from 10 to 20 m (Figure 3.5b) can be written as follows:

$$Q_1'C_1' + Q_{g2}'C_g = Q_2'C_2' \tag{3.9}$$

$$Q_1' + Q_{g2}' = Q_2' \tag{3.10}$$

Where the Q'_{g2} and Q'_2 are the groundwater seepage rates in the segment and the discharge at the end of the segment in $m^3/$; C_g and C'_2 are the EC values of the groundwater seepage and stream water at the end of the segment in μS/cm, respectively. With the estimated Q'_1 in Equations 3.7 and 3.8, Q'_{g2} and Q'_2 can be computed with the formula:

$$Q'_2 = Q'_1 \frac{C'_1 - C_g}{C'_2 - C_g} \tag{3.11}$$

$$Q'_{g2} = Q'_2 - Q'_1 \tag{3.12}$$

The stream discharge and groundwater seepage at the remaining segments can be calculated consecutively with Equations (3.11) and (3.12).

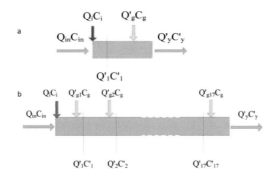

Figure 3.5 Schematic plot of mass balance calculations under constant injection: (a) for the total 180-m reach; and (b) for 10-m segments.

3.2.6.4 Estimation of seepage using the EC profile under a constant injection

Stream discharge and groundwater seepage can be calculated by using both the natural EC profile and the EC values under the constant injection. The mass balance equations for the total 170-m reach are:

$$Q_1 C_1 + Q_g C_g = Q_y C_y \tag{3.13}$$

$$Q'_1 C'_1 + Q'_g C_g = Q'_y C'_y \tag{3.14}$$

$$Q_1 + Q_g = Q_y \tag{3.15}$$

$$Q'_1 + Q'_g = Q'_y \tag{3.16}$$

Equations (3.13) - (3.16) can be solved for Q_1 and Q_g when assuming that the stream discharge and groundwater seepage remain the same during the natural and constant injection EC measurements. Indeed, the difference of stream discharge at the Yujiawan gauging station is very small between the natural and constant injection EC measurements. The average discharge at the Yujiawan from 10:00 to 13:00 can be used and is calculated to be 0.0314 m³/s. Therefore, Q_1 and Q_g along the 170-m river reach can be calculated. The mass balance equations for the second 10-m segment from 20 to 30 m downstream under the natural and constant injection conditions are:

$$Q_1 C_1 + Q_{g2} C_g = Q_2 C_2 \tag{3.17}$$

$$Q_1' C_1' + Q_{g2}' C_g = Q_2' C_2' \tag{3.18}$$

Equations (3.17) and (3.18) can be solved to calculate Q_2 and Q_{g2}. Stream discharge and groundwater seepage at the remaining segments can be calculated successively.

3.2.6.5 Sensitivity analysis

The reliability of tracer methods for estimating groundwater and surface-water exchange was evaluated by Ge and Boufadel (2006) and Wagner and Harvey (2001). They concluded that the stream tracer approach had minimal sensitivity to the surface-subsurface exchange at high baseflow conditions. As indicated by the hydraulic gradients and thermal methods, the groundwater discharge to the surface water dominates the main interaction between groundwater and surface water in this 180 m reach, which reduces the probability of overestimation of groundwater discharge caused by losing solute in the process. For the seepage calculation methods in this case study, uncertainties may result from the estimation error of the groundwater EC value along the stream bank, EC measurement errors along the reach, and discharge measurement error at the gauging station. Given unknown distribution of the error distribution of the calculation, a simple sensitivity analysis was carried out to investigate relative errors in seepage calculation caused by the likely EC and discharge measurement errors. The range of the EC values from the groundwater monitoring wells in the stream valley is 510 μS/cm, which is very large and most likely caused by irrigation. Groundwater EC values along the stream bank are expected to be lower than the measured EC values of natural stream water. Therefore, a 5% variation with respect to the average value is assumed for the sensitivity analysis. The sensitivity of EC measurement errors on the calculation of groundwater seepage and discharge was investigated by systematically increasing and decreasing of EC values along the river by 5%.

3.3 Results and Discussion

3.3.1 Measurement and analysis

3.3.1.1 River discharge measurement

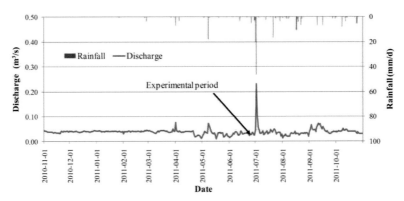

Figure 3.6 Stream discharge at Yujiawan gauging station and rainfall at the rain gauge from 1 November 2010 to 31 October 2011.

Several peaks corresponding with rainfall events can be observed in Figure 3.6. The maximum daily discharge is 0.231 m^3/s on 2 July 2011 when a heavy rainfall event occurred. The minimum daily discharge is 0.0105 m^3/s on 17 May 2011, which was caused by water diversions and groundwater abstraction for irrigation from the end of April till the beginning of October. The stream flow keeps relatively stable during the winter, since there is hardly any precipitation and no anthropogenic water use during this period. Furthermore, the constant discharge indicates groundwater sustaining the base flow. The average discharge on 21 June 2011was 0.0316 m^3/s from 10:00 to 11:00, and 0.0321 m^3/s from 13:00 to 14:00 when the constant injection experiment was carried out with an injection rate of 89.7 ml/min. The discharge increase after 18:00 was caused by the stopping of water diversions and groundwater abstraction for irrigation in the adjacent flood plain.

3.3.1.2 Groundwater measurement

The groundwater levels in boreholes (Figure 3.1) at the terrace (well a), the flood plain (well b), stream bank (well c), and below the streambed of the Yujiawan stream are shown in Figure 3.7. Groundwater levels are relatively stable through the year. Heavy rainfall events in July cause an abrupt rise of the groundwater levels in all wells.

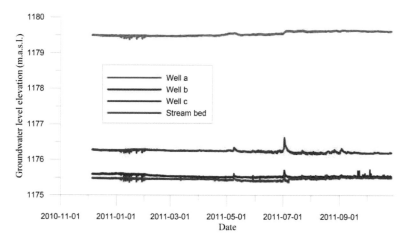

Figure 3.7 Groundwater levels below the terrace (Well a), flood plain (Well b), stream bank (Well c), and streambed from 1 November 2010 to 31 October 2011.

3.3.1.3 Temperature measurements

Figure 3.8 shows the temperature measurements at the streambed surface and at different depths beneath the streambed from September 2010 to October 2011. The gap in April was caused by the limitation of data storage capacity of the data loggers. The temperature at a depth of 80 cm beneath the streambed was constant throughout the year. The temperature at depths of 50 and 30 cm show some small fluctuation, while the temperature at the surface of the streambed shows large daily and seasonal variations.

3.3.1.4 Selected hydrochemical parameters

Figure 3.9 illustrates changes of stable isotope values for deuterium and oxygen-18, and the concentration changes of the cations potassium (K^+) and sodium (Na^+) as well as the anions chloride (Cl^-) and nitrate (NO_3^-) in river water samples from 1 to 5 July 2011 , which included the heavy rainfall on 2 July. After the heavy rainfall event on 2 July, stable isotope values decreased, while concentrations of both cations and anions increased. The concentrations of the cations (K^+ and Na^+) and anions (Cl^- and NO_3^-) in stream flow samples increased during the discharge event caused by the heavy rainfall event at the beginning of July (Figure 3. 9).

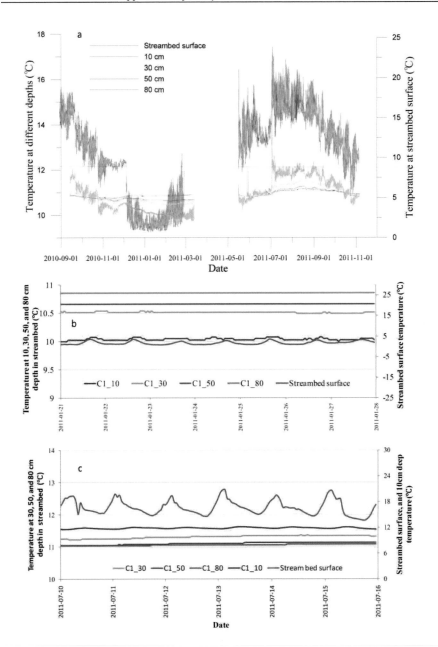

Figure 3.8 (a) Temperature of stream water and stream sediment at 10, 30, 50, and 80-cm depth beneath the stream bed from 1 September 2010 to 31 October 2011. Temperature at the surface of the streambed (right y axis) and at different depths beneath the streambed in (b) winter and (c) summer in 2011.

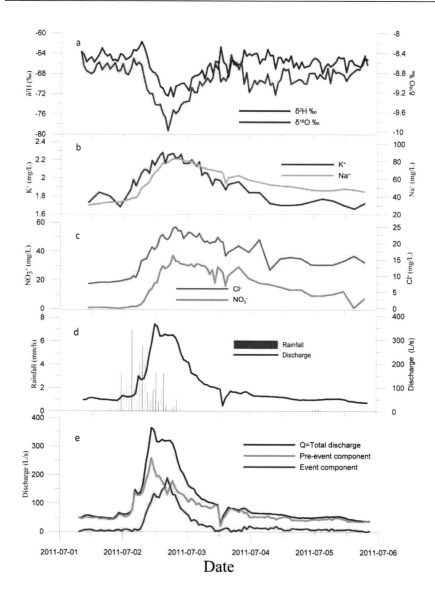

Figure 3.9 (a, b, c) Stable isotope values, hydrochemical behavior of streamflow samples, (d) rainfall and discharge from 1 July to 5 July 2011, and (e) a two-component hydrograph separation using oxygen-18 for the rainfall event that occurred on 2 July 2011.

3.3.1.5 Constant injection tracer experiment

The electrical conductivity (EC) values along the Yujiawan stream from the constant injection point to the discharge gauging station located 180 m downstream were measured for the natural status (before conducting the constant injection experiment) and under the constant injection condition. Figure 3.10 illustrates the difference in EC values along the Yujiawan stream for natural and injection conditions. The EC values for natural stream water vary from 640 µS/cm at the constant injection point to 581 µS/cm at 180 meter downstream at the gauging station. The highest EC value of 882 µS/cm was measured 10 m downstream of the injection point during the constant injection condition and the EC values decreased gradually to 660 µS/cm at 180 m downstream.

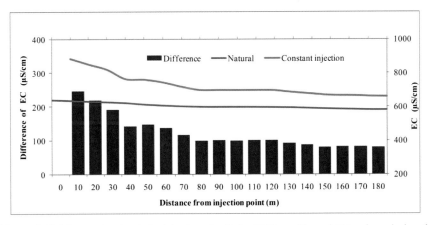

Figure 3.10 Measured natural electrical conductivity (EC) profile and EC values during the constant injection experiment on 21 June 2011.

3.3.2 Discussion

3.3.2.1 Hydraulic method

The typical distribution of groundwater levels in the sub-catchment at 20:00 14 June 2011 is plotted in Figure 3.7. It indicates that groundwater flows from both hill slopes towards the stream. Groundwater discharge to the stream occurs during the whole year since groundwater levels in the valley are always higher than the stream stage during the period from December 2010 to October 2011 (Figure 3.7).

3.3.2.2 Temperature method

For a gaining stream reach with upward advection, the oscillating surface temperature signal is attenuated at shallow depths, such that the greater the discharge, the greater the attenuation of temperature extremes, and the greater the lag in temperature extremes in the sediments. Temperature measurements (Figure 3.8) also show groundwater discharge to the stream for the whole investigated period.

Variations of temperatures during a week in winter (21-28 January 2011) and in summer (10-16 July 2011) are plotted to view temperature differences at different depths. During winter (January), the air temperature is below zero, and the stream-water temperature at the stream bottom is still above the freezing point with small diurnal fluctuations (Figure 3.8b). Temperature increases with the increase in depth, and the temperature at 80 cm remains constant at around 11°C, which represents groundwater temperature. These conditions show that both convective and dispersive heat transport are from groundwater upward to the stream. The higher temperature of the groundwater prevents stream water from freezing. In the summer period (July), the stream-water temperature at the stream bottom is much higher than the groundwater temperature (Figure 3.8c). The large diurnal fluctuations of stream water temperature do not penetrate into the streambed deposits. However, temperature decreases with the increase of depth. The lack of diurnal fluctuations in sediment temperature indicates that convective heat transport is from the groundwater upward to the stream, while dispersive heat transport is from the stream downward to the groundwater.

3.3.2.3 Hydrograph separation

The increased concentrations of chemical components after the start of the rainfall event could be caused by the dissolution of salts which have accumulated in the top soil of the agricultural fields or from the fertilizers used for agriculture in the sub-catchment. Discharge suddenly decreased from 3 July, as did the concentrations of several chemical constituents, which could be due to the decrease of event-water with high chemical constituents discharged into the stream after the stop of rainfall on 2 July. It was not possible to carry out hydrograph separations using these tracers, because it was not possible to measure the representative end-member concentration of the runoff components or to assume that they are constant in time. Thus, the stable isotope oxygen-18 was used for a two-component hydrograph separation to separate pre-event and event water (Figure 3.9). The groundwater discharge to the stream could be estimated through the proportion of pre-event water to the total stream flow while the event water represents the surface runoff to the stream. Light rainfall occurred at the end of June, followed by the heavy event starting from 1 July. The event samples were collected from 1 July till 5 July. There was intensive rainfall during the 12 h on 2 July within the Bulang sub-catchment. The isotopic components in the rainfall and groundwater were assumed to be constant in space and time of the duration of the investigated event. The results of hydrograph separation illustrate that the pre-event component accounts for 74.8% of the total discharge during this heavy rainfall event. The response of the pre-event component to the rainfall is faster than the event water component that might be caused by the so-called kinematic wave effect (Buttle, 1994, Cey, et al., 1998, Krein and De Sutter, 2001)), which is caused by a faster flood wave propagation velocity compared to the flow velocity of water. This results in a high groundwater discharge component in the stream also during discharge events.

3.3.2.4 Seepage calculation with EC profile

Calculated results

Figure 3.11 and Table 3.1 present the results of seepage estimation by the three methods. For a relatively homogeneous sandy aquifer like in this test case, groundwater seepage along this river reach shows large spatial variations that might be caused by the variations of the streambed topography. Large spatial variations of groundwater discharge to rivers were also found in other studies using temperature surveys (Becker, et al., 2004, Lowry, et al., 2007). It can be seen that the groundwater seepages calculated by the constant injection and combined method are very close (Figure 11). The natural EC profile method calculates slightly lower seepage rates. All three methods calculate a comparable discharge value at the gauging station (Table 3.1). The calculated seepage under a bridge (at 100 m distance) is zero since the concrete wall of the bridge is not permeable. Groundwater seepage is not uniform along the reach. In segments with very low seepage rates, the combined method calculated negative seepage rates which are possibly caused by local-scale streambed topography variations or by measurement errors of the EC values. The application of natural and constant injection methods requires the estimation of EC values of groundwater along the stream bank. The combined method eliminates the unknown groundwater EC values, therefore, it is more convenient to use.

Table 3.1 Seepage calculation for the 180-m reach by the three EC-profile methods

	Natural	Constant injection	Combined
Inflow from upstream (m^3/s)	0.01197	0.01095	0.01009
Average seepage (m^3/s)	0.0196	0.02115	0.02133

Discharge at 180-m downstream (m³/s)	0.0316	0.0321	0.0314
Groundwater seepage rate (m³/m/d)	9.41	10.75	10.84

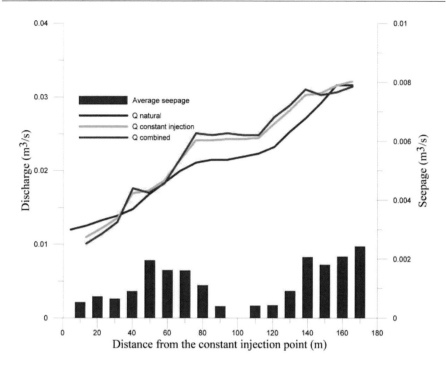

Figure 3.11 Estimation of seepage and discharge along the stream reach.

Sensitivity analysis

The results of sensitivity analysis are shown in Figure 3.12. It is clear that the natural EC profile method is also more sensitive to EC measurement errors. The measured mean discharge at the gauging station from 10:00 to 14:00 was used for the estimation of the groundwater seepage with the combined method. The standard deviation of the discharge during the experiment is 0.00143 m³/s. Thus, the sensitivity analysis to the discharge measurement errors was analyzed by increasing and decreasing the discharge with two standard deviations. Figure 3.13 shows that the sensitivity of the estimated river discharge along the reach to the discharge measurements at the downstream gauge station is smaller than the likely measurement error at the gauging station.

The sensitivity analysis indicates that the combined method for determining groundwater seepage provides the most reliable estimations. Another advantage of the combined method is that it does not need to measure EC values of groundwater along the stream bank, therefore, reducing measurement costs.

The hydraulic method indicates the interaction between the groundwater and surface water by providing the difference between groundwater heads and river stage. The temperature method could be applied in losing or gaining reaches at carefully selected locations, which identifies the direction of water flow in the streambed, However, either the hydraulic method or temperature method should only be employed for quantitative estimation with external hydraulic information such as the hydraulic conductivity, and head difference between groundwater and river stage. Hydrograph separation could only be applied during rainfall events with extensively field work, isotopic and chemical analysis in the laboratory. The EC profile method can be finished within hours, which is more efficient compared with other methods.

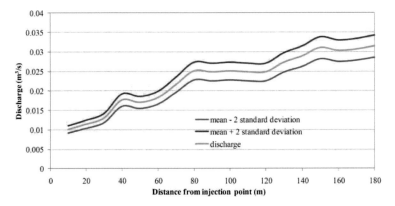

Figure 3.12 Sensitivity of estimated stream discharge to groundwater EC value along the stream bank (a) with the natural EC profile method, (b) with the constant injection method. Sensitivity of estimated stream discharge to EC measurement errors (c) with the natural EC profile method, (d) with the constant injection method, and (e, f) with the combined method.

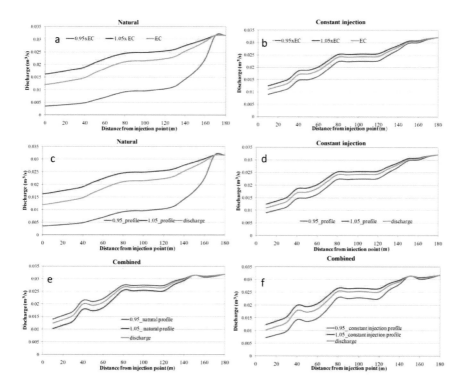

Figure 3.13 Sensitivity of estimated stream discharge along the reach to the discharge measurement error at the gauging station using the combined method.

3.4 Conclusion

Multiple field measurements for an investigation period of 1 year indicate that groundwater discharges to the river during the entire period in the Bulang sub-catchment. Even during heavy rainfall events, river discharge is composed of more groundwater discharge than direct surface runoff, which is in line with other field observations and the high infiltration rates in the catchment (dominated by sand dunes). Consequently, groundwater and stream water are essentially one resource and need to be managed conjunctively.

Groundwater level measurements in a cross-section in relation to stream stage measurements provide indication of flow directions of groundwater and stream interactions. The measurements in the Bulang sub-catchment show that groundwater levels are higher than the stream stage for the whole measurement period, a clear evidence of groundwater discharge to the stream ear round. The temperature

57

measurements provide additional information on groundwater and stream interactions. Groundwater has a relatively constant temperature, while stream-water temperature has not only seasonal changes, but also diurnal fluctuations. Temperature measurements in the streambed deposits at different depths can identify the direction of water exchange between groundwater and surface water. The Bulang River never freezes despite the cold temperature in the winter, due to groundwater seepage with a higher temperature. Both convective and dispersive heat transports occur in the same upward direction from groundwater to the stream in the winter period. In summer, stream-water temperature is much higher than groundwater temperature. The dispersive heat transport is downward, but the convective heat transport is upward. The large diurnal fluctuation of stream-water temperature is dammed by the upward cold groundwater flow. Therefore, temperature of streambed deposits below 30 cm depth remains stable, while the temperature at shallow depth shows small fluctuations.

River gauging is indispensable for quantifying groundwater and stream interactions. First, hydrograph separation of measured stream discharge provided an estimate of total groundwater discharge to the stream. Second, mass balance equations with EC measurements were solved more accurately with the measured discharge at the gauging station. The combined use of EC profile measurements under the natural situation with the constant injection can result in spatially distributed groundwater seepage estimates along the stream. Sensitivity analysis demonstrates that the combined method is neither very sensitive to the EC measurement errors, nor to the measurement errors of the discharge measurements, and provides the most reliable estimation of groundwater seepage.

4. Groundwater-Surface Water Interactions under Different Land Use Scenarios[*]

Abstract: Groundwater is the most important resource for local society and the ecosystem in the semi-arid Hailiutu River catchment. The catchment water balance was analyzed by considering vegetation types with the Normalized Difference Vegetation Index (NDVI), determining evapotranspiration rates by combining sap flow measurements and NDVI values, recorded precipitation, measured river discharge and groundwater levels from November 2010 to October 2011. A simple water balance computation, a steady state groundwater flow model, and a transient groundwater flow model were used to assess water balance changes under different land use scenarios. It was shown that 91% of the precipitation is consumed by the crops, bushes and trees; only 9% of the annual precipitation becomes net groundwater recharge which maintains a stable stream discharge in observed year. Four land use scenarios were formulated for assessing the impacts of land use changes on the catchment water balance, the river discharge, and groundwater storage in the Bulang catchment. The scenarios are: (1) the quasi natural state of the vegetation covered by desert grasses; (2) the current land use/vegetation types; (3) the change of crop types to dry resistant crops; and (4) the ideal land use covered by dry resistant crops and desert grasses, These four scenarios were simulated and compared with measured data from 2011, which was a dry year. Furthermore, the scenarios (2) and (4) were evaluated under normal and wet conditions for years in 2009 and 2014, respectively. The simulation results show that replacing current vegetation and crop types with dry resistant types can significantly increase net groundwater recharge which leads to the increase of groundwater storage and river discharges. The depleted groundwater storage during the dry year could be restored during the normal and wet years so that groundwater provides a reliable resource to sustain river discharge and the dependent vegetations in the area.

---

[*].This chapter is based on the paper Simulation of Groundwater-Surface Water Interactions under Different Land Use Scenarios in the Bulang Catchment, Northwest China. Yang, Z., Zhou, Y., Wenninger, J., Uhlenbrook, S., & Wan, L. (2015). Water, 7(11), 5959-5985. DOI:10.3390/w7115959.

4.1 Introduction

Water is the most important limiting factor for agricultural production and ecosystem protection in semi-arid conditions (Snyman and Fouché, 1991). To maintain a delicate balance between the protection of the ecosystem and the sustainable development of local society is critical task in semi-arid regions. Optimizing water use efficiency for agricultural crops (Deng, et al., 2006) is a key approach to mitigate water shortages and to reduce environmental problems in arid and semi-arid regions. Along with the water shortage limitation for agricultural productivity, the desertification has been controlled by planting shrubs as ecosystem rehabilitation measures in recent decades in the semi-arid parts of northwest China (Mitchell, et al., 1998). However, the water balance in semi-arid catchments can be significantly influenced by vegetation type (Bellot, et al., 1999), irrigation schedules (Mermoud, et al., 2005), and groundwater irrigation (Mermoud, et al., 2005). Brown et al. (2005) determined that the changes in water yield at various time scales were a result of permanent changes in the vegetation cover by means of mean annual water balance model. Jothityangkoon et al. (2001) concluded that spatial variability of soil depths appears to be the most important controlling parameter for runoff variability at all time and space scales, followed by the spatial variability of climate and vegetation cover in semi-arid catchments. The differences in water balance components between a number of temperate and semi-arid catchments in Australia can be attributed to the variability of soil profile characteristics like water storage capacity and permeability, vegetation coverage and water use efficiency, rainfall, and potential evaporation (Farmer, et al., 2003). Scott et al. (2000) indicated that grassland relies primarily on recent precipitation, while the tree/shrub obtained water from deeper parts of the soil profile in the semi-arid riparian floodplain of the San Pedro River in southeastern Arizona. Planting of shrub seedlings can significantly enhance topsoil development on the dune surface and stabilizing the sand dunes (Li, et al., 2009), but the water consumption of artificially introduced plants and the effects on the water balance have not been investigated.

Land use management and rehabilitation strategies would have significant impact on the catchment water balance and hence on water yield and groundwater recharge (Zhang, et al., 2001). Consequently, understanding impacts of land use change on the

60

catchment water balance dynamics is critical for sustainable water resources management. Stream flow (Lørup, et al., 1998, Bronstert, et al., 2002, Niehoff, et al., 2002), flow regime (Yang, et al., 2012) and the river water quality (Ahearn, et al., 2005, Zampella, et al., 2007) can be influenced by the change of land use/land cover in the catchment. Compared with the mechanism of land use effects on the surface water, the groundwater recharge (Harbor, 1994, Scanlon, et al., 2005), discharge (Salama, et al., 1999, Batelaan, et al., 2003), levels (Furukawa, et al., 2005), hydrochemistry and contamination (Jeong, 2001), and nitrate concentrations (McLay, et al., 2001, Molénat and Gascuel‐Odoux, 2002, Cole, et al., 2006, Choi, et al., 2007) can be indirectly affected by the land use changes through infiltration. Krause et al. (2007) assessed the impacts of different strategies for managing wetland water resources and groundwater dynamics of landscapes based on the analysis of model simulation results of complex scenarios for land-use changes and changes of the density of the drainage-network, but regional groundwater modeling studies are often hampered by data scarcity in space and time especially in semi-arid regions (Leblanc, et al., 2007). Thus, the estimation of areal inputs such as precipitation and actual evapotranspiration (ET) is essential for groundwater model studies. Actual ET corresponds to the real water consumption and is usually estimated by considering weather parameters, crop factors, management and environmental conditions (Courault, et al., 2005). Despite of some uncertainties and inconsistencies in the results, remote sensing is a useful technique for the study of groundwater hydrology and has aided the successful location of important groundwater resources (Waters, et al., 1990). Remote sensing and Geographic Information Systems (GIS) have been used for the investigation of springs (Sener, et al., 2005), determining the groundwater dependent ecosystems (Münch and Conrad, 2007), determining the recharge potential zones (Shaban, et al., 2006), mapping groundwater recharge and discharge areas (Tweed, et al., 2007), detecting potential groundwater flow systems (Bobba, et al., 1992), and monitoring infiltration rates in semi arid soils (Ben-Dor, et al., 2004). Remote sensing data can be employed for estimating the ET by means of energy balance methods, statistical methods using the difference between surface and air temperature, surface energy balance models, and spatial variability methods at different scales (Kalma, et al., 2008). Remote sensing and the Normalized Difference Vegetation Index (NDVI) have been widely employed for estimating groundwater evapotranspiration (Groeneveld, et al., 2007, Groeneveld, 2008), investigating the relationship between vegetation growth and depth to groundwater table (Jin, et al., 2007, Jin, et al., 2011, Lv, et al., 2013, Zhou, et al., 2013), and accessing groundwater recharge fluxes (W Lubczynski and Gurwin, 2005). Many researchers have conducted direct measurements like the sap flow method which uses the stem heat balance technique (Bethenod, et al., 2000), and scaled these values to the catchment level transpiration (Ford, et al., 2007). The remote sensing could offer the relevant spatial data and parameters at the appropriate scale for use in distributed hydrological models (Stisen, et al., 2008) and groundwater models (Hendricks Franssen, et al., 2008, Li, et al., 2009). Those studies trend to utilize the areal ET calculated from the theory of surface energy balance with remote sensing data at lager scale. However, few studies have been conducted on simulating the groundwater response to different land use scenarios by determining areal ET using field measurements and remote sensing data in semi-arid regions particularly in Asia.

Most recent studies focus on the simulation of land use or climate change impacts for improving the water use efficiency for agricultural production, but few have been conducted on artificially introduced plants for stabilizing sand dunes and the effects on the water resources in semi-arid regions. Since evapotranspiration from crops and shrubs dominate groundwater discharge in semi-arid catchments (Yang, et al., 2014), a balance must be achieved between land use (prevention of desertification) and water resources conservation for agricultural and other purposes. Sap flow measurements can directly provide transpiration rates of the individual vegetation in carefully selected sites. However, the uncertainty of estimating areal ET rates varies in space and time through the necessary up-scaling for the areal computation. Remote sensing technologies have been recently employed for calculating the areal ET by means of energy balance at large scale, but few were conducted using combined field measurements and remote sensing data (NDVI) in semi-arid regions. Multiple procedures have been employed in this study in order to evaluate the influence of land use management on hydrological processes in the semiarid Bulang catchment. The methods used in this study consisted of identifying vegetation types with NDVI values, determining evapotranspiration rates with sap flow measurements in the field, and computing the areal ET rates by a combination of remote sensing data (NDVI) and field measurements. Impacts of different land use scenarios on water resources were simulated by a groundwater model with the aid of combined field measurements, remote sensing, and GIS techniques, which reveals the catchment water balance in the sandy region of the middle section of the Yellow River Basin. The results provide scientific information to support rational land use management and water resources conservation in semi-arid areas.

4.2 Materials and Methods

4.2.1 Study area

The Bulang River is a tributary of the Hailiutu River, which is located in the middle reach of the Yellow River in northwest China (Figure 4.1). Located in a semi-arid region, the total area of the Bulang catchment is 91.7 km^2. The land surface is characterized by undulating sand dunes, scattered desert bushes (*Salix Psammophila*), cultivated croplands (*Zea mays*), and a perennial river in the southwestern downstream area. The surface elevation of the Bulang catchment ranges from 1300 m at the northeastern boundary to 1160 m above mean sea level at the catchment outlet in the southwest. The long-term annual average daily mean temperature is 8.1 °C and the monthly mean daily air temperature is below zero in the winter time from November until March. The mean annual precipitation measured at the nearby Wushenqi meteorological station for the period 1984 to 2011 was 340 mm/year. Precipitation mainly falls in June, July, August and September. The mean annual pan evaporation (recorded with an evaporation pan with a diameter of 20 cm) is 2184 mm/year (Wushenqi metrological station, 1985–2004). The geological formations in the Bulang catchment mainly consist of four strata (1) the Holocene Maowusu sand dunes with a thickness from 0 to 30 m; (2) the upper Pleistocene Shalawusu formation (semi consolidated sandstone) with a thickness of 5 to ~90 m; (3) the Cretaceous Luohe sandstone with a thickness of 180 to ~330 m; and (4) the underlying impermeable Jurassic sedimentary formation.

4.2.2 Methods

4.2.2.1 Water balance

The catchment water balance can be calculated with the components of precipitation (P), evapotranspiration (ET), discharge of the Bulang stream (Q), deep groundwater circulation discharging to the main Hailiutu River (D), and the change of storage (dS/dt) in the catchment as:

$$P - ET - Q - D = \frac{ds}{dt} \tag{4.1}$$

Precipitation, discharge of the Bulang River, and the change of the groundwater levels that represent the storage change in the system can be directly measured in the field, but not the ET and deep circulation fluxes. In this study, the areal ET was estimated from the site measurements of sap flow of maize, salix bush, and willow tree which were up scaled by using the NDVI generated vegetation cover from remote sensing data.

4.2.2.2 Groundwater discharge

The groundwater discharge as the baseflow was separated from daily average river discharge measurements according to hydrograph separation results using isotopic tracers (Yang, et al., 2014). Daily river discharge can be considered as groundwater discharge during days without rainfall, while 74.8% of the river discharge originated from the groundwater in the two days after the recorded heavy rainfall based on the trace method (Yang, et al., 2014).

4.2.2.3 Groundwater model

In correspondence with the geological structure, the upper Maowusu sand dunes together with Shalawush deposits were simulated with one model layer as the unconfined aquifer and the underlying Cretaceous Luohe sandstone was simulated with another model layer as the confined aquifer.

The outlet of the river valley was simulated as a head-dependent flow boundary, considered to be the deep groundwater discharge to the Hailiutu River. All other boundaries were taken as no-flow boundaries in line with the catchment water divide. The Bulang stream was simulated as a drain since groundwater always discharges to the stream. Net groundwater recharge was calculated as precipitation (P) minus evapotranspiration (ET).

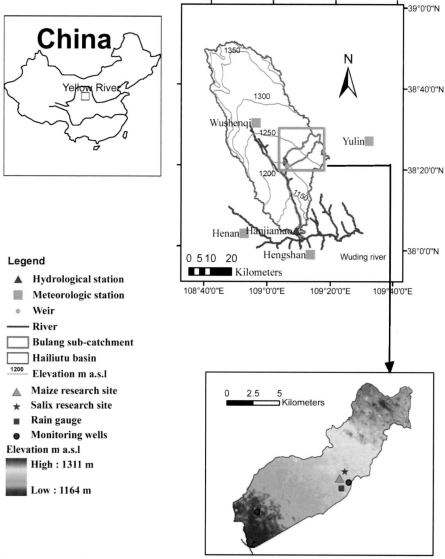

Figure 4.1 Location of the Bulang sub-catchment within the Hailiutu basin in Northwest China, the location of sap flow measurements for maize and salix, hydrological and meteorological stations, groundwater level monitoring wells, and the digital elevation model.

The popular numerical model code MODFLOW (McDonald and Harbaugh, 1984) was used for the flow simulation. The numerical model grid cell has a uniform size of 50 by 50 m, resulting in a model grid consisting of 310 rows and 350 columns. The top elevation of the model is the surface elevation which is taken from the Digital Elevation Model with a resolution of 30 by 30 m. The bottom elevation of the model is interpolated from the limited geological borehole data. The hydrogeological

65

parameters such as hydraulic conductivity, specific yield, and specific storage were divided into three zones according to lithology. The General-Head Boundary (GHB) package of MODFLOW was used to simulate deep groundwater discharge to the Hailiutu River, where flow into or out of a GHB cell is calculated in proportion to the difference between the head in the model cell and the stage of the Hailiutu River. The MODFLOW Drain package was used to simulate groundwater discharge to the Bulang stream since the stream acts as a drain which always receives groundwater discharge. The Recharge package was used to simulate the net recharge which was calculated in an Excel sheet being precipitation minus evapotranspiration.

A steady state flow model was calibrated first with the annual average values of rainfall, ET rates, stream discharges, and the groundwater levels measured from November 2010 to October 2011. A transient flow model was then constructed and calibrated with the measured daily data from November 2010 to October 2011. The model calibration was performed with the optimization code PEST (Doherty, et al., 1994). PEST found optimal values of hydraulic conductivities and storage parameters by minimizing the sum of squared differences between model-calculated and observed values of groundwater heads at the observation wells. The Drain conductance was further calibrated by comparing calculated and measured stream discharges.

4.2.2.4 Scenario analysis

For semi-arid areas such as our study area, a delicate balance among vegetating sand dunes, agricultural production, and water resources conservation must be maintained (Zhou, et al., 2013). Over-vegetating sand dunes with high water consumption species (such as poplar trees and salix bushes) would reduce net groundwater recharge and diminish stream flows. Desert grasses, such as *Artemisia Ordosica* and *Korshinsk Peashrub*, consume much less water and are better choice for vegetating sand dunes.

These grasses are common in areas with a deep groundwater table. The dominant agricultural crop type in the area is maize. Maize needs to be irrigated six times (Hou, et al., 2014) and consumes a lot of water. The local agricultural department has started a program to extend dry-resistant crops such as sorghum and millet. Considering the possible land use changes in the area, the following scenarios are proposed:

- Scenario 1 assumes a natural situation in which the catchment is covered by desert grasses. This scenario sets up a bench mark case to compare impacts of land use changes on groundwater and stream flow.

- Scenario 2 represents the current land use in 2011. Impacts of current land use changes on groundwater and stream flow can be analyzed by a comparison to scenario 1.

- Scenario 3 simulates effects of replacing maize crops with less water consumptive crops such as sorghum and millet in order to provide support to the policy of extending dry-resistant crops in the area.

- Scenario 4 proposes an ideal land use scenario where sand dunes are covered by desert grasses and dry-resistant crops are grown. This scenario gives direction to the future land use changes in the area.

Since there are inter-annual variations of precipitation and evapotranspiration accordingly, scenarios 2 and 4 were simulated under the dry, normal, and wet years. Analysis of long-term annual precipitation (Wushenqi meteorological station, 1959–2014) found that 2011, 2009, and 2014 represent dry (87.7%), normal (50.2%), and wet (27.8%) years, respectively. These two land use scenarios were further simulated with a multiple year transient groundwater model in order to assess impacts of inter-annual variations of climate.

4.2.2.5 Model inputs for the simulation of land use scenarios

The calibrated transient groundwater flow model was used to simulate these four land use scenarios. Given inputs of 8-days net recharge for November 2010 to October 2011, the model simulates groundwater level changes and computes groundwater discharge to the Bulang River, deep groundwater circulation to the Hailiutu River and change of groundwater storage.

The net recharge was calculated to be the precipitation minus evapotranspiration. The evapotranspiration was estimated in the same procedure as before with equations of:

$$ET_i = \frac{NDVI_i}{NDVI_{plant}} ET_{plant} \quad , \quad 0 < NDVI_i < 0.4 \tag{4.2}$$

$$ET_i = \frac{NDVI_i}{NDVI_{crop}} ET_{crop} \quad , \quad NDVI_i \geqslant 0.4 \tag{4.3}$$

Where ET_i is estimated model grid ET value. ET_{plant} and ET_{crop} were found from relevant studies (Dong, et al., 1997) and are presented in Table 4.1.

Table 4.1 ET values for reference crops and plants for scenario simulations (mm/year).

Scenario	Rainfall	Crop ET	Plant ET	NDVI Map
1-Natural situation	214.8	*Artemisia Ordosica* (133.6)	*Artemisia Ordosica* (133.6)	August 2011
2-Current land use	214.8	Maize (503.1)	*Salix Psammophila* (245.1)	August 2011
3-Dry-resistant crop	214.8	Sorghum (402.5)	*Salix Psammophila* (245.1)	August 2011
4-Ideal land use	214.8	Sorghum (402.5)	*Artemisia Ordosica* (133.6)	August 2011

In order to evaluate impacts of inter-annual variations of precipitation on the water balances, scenarios 2 and 4 were further simulated under a normal and wet years. The year of 2009 represents a normal year and 2014 represents a wet year. Daily precipitation of Wushenqi station was collected to compute net recharge. NDVI maps of 2009 and 2014 were processed to estimate ET values. These data are presented in Table 4.2.

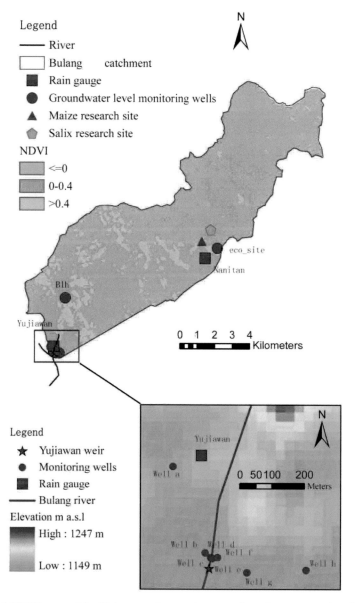

Figure 4.2 NDVI map of the Bulang catchment, interpretation of remote sensing data from (TM) image, observed on August 2011.

4.2.3 Field Measurements

4.2.3.1 Precipitation and Evapotranspiration

In order to measure the hydrological variables, a set of instruments had been installed in the catchment. Precipitation was measured from November 2010 to

October 2011 using two Hobo rain gauges (RG3-M Data Logging Rain Gauge, Onset Corporation, Bourne, MA, USA) at the outlet of the catchment and at Nanitan (Figures 4.1 and 4.2). Stem flow sensors (Flow 32, Dynamax, Houston, TX, USA) were used to measure sap flow in maize stems at the maize research site (Hou, et al., 2014) and in salix stems at the salix research site (Huang, et al., 2014) during the growing season in 2011. The NDVI map (resolution 30 m × 30 m) generated from remote sensing data have been utilized for classifying the land use and vegetation types in the area (Figure 4.2). NDVI values larger than 0.4 indicate crop land, between 0 and 0.4 are desert bushes, and smaller than 0 are considered as bare sand.

Table 4.2 Precipitation and ET values for reference crops and plants for scenarios 2 and 4 (mm/year).

Scenario	Rainfall	Crop ET	Plant ET	NDVI Map
2-Current land use	Dry (214.8) Normal (340.0) Wet (420.0)	Maize (503.1)	*Salix Psammophila (245.1)*	August 2011 August 2009 September 2014
4-Ideal land use	Dry (214.8) Normal (340.0) Wet (420.0)	Sorghum (402.5)	*Artemisia Ordosica* (133.6)	August 2011 August 2009 September 2014

4.2.3.2 Discharge gauge

One discharge gauging station was constructed at the outlet of the Bulang catchment (Figure 4.1). The Yujiawan gauging station consists of one permanent rectangular weir equipped with an e + WATER L water level logger (Type 11.41.54, Eijkelkamp Agrisearch Equipment, Giesbeek, The Netherlands) where water levels are recorded with a frequency of 30 min. Water depths are converted to discharges using a rating curve based on regular manual discharge measurements carried out with a current meter and the velocity-area method.

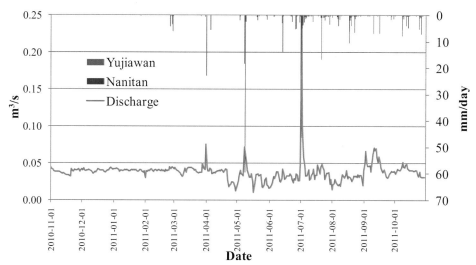

Figure 4.3 Rainfall at Yujiawan and Nanitan station, measured discharge at Yujiawan weir from November 2010 to October 2011.

4.2.3.3 Observation wells

Several groundwater level monitoring wells were installed in the Bulang catchment (Figures 4.1 and 4.2). Equipped with submersible pressure transducers (Type: MiniDiver, Eijkelkamp Agrisearch Equipment, Giesbeek, The Netherlands), the groundwater levels in these observation wells were recorded at 10-min intervals. Barometric compensation was carried out using air pressure measurements from a pressure transducer (Type: BaroDiver, Eijkelkamp Agrisearch Equipment, Giesbeek, The Netherlands) installed at the site. Groundwater levels were converted into the height of the water table above mean sea level with the calibration of the land elevation of wells, height of water column above the MiniDiver in the wells, and the depth of the MiniDiver in the boreholes.

4.3 Results

4.3.1 Results

4.3.1.1 Estimation of Catchment water balance

The total observed precipitation from November 2010 to October 2011 at Yujiawan and Nanitan station are 206.8 mm and 217.6 mm, respectively. The annual areal precipitation in the catchment was estimated to be 214.8 mm/year by means of area weighed average of two rainfall stations. Discharge at the Yujiawan weir varies from 0.01 to 0.23 m^3/s (Figure 4.3). Stream discharge is very stable during the winter months and varies in the summer months due to irrigation water use and sporadic rainfall. A heavy rain event occurred on 2 July and generated the highest discharge in the observation period.

ET rates measured at the maize and salix research site were 503.1 mm/year and 245.1 mm/year, respectively. The NDVI values at the sap flow measurement sites for maize and salix are 0.59 and 0.17, respectively. The NDVI was used to upscale evapotranspiration rate measured in the salix and maize research sites to the catchment using Equations (4.2) and (4.3).

The areal ET value of the catchment was calculated as the average ET of all cell ET values. Figure 4.4a illustrates the up-scaling approach for areal ET with sap flow measurements and NDVI values. Figure 4.4b plots the estimated daily ET rates. The daily ET rates increases from mid April, peak in July and August, and decreases from September onwards. Daily ET rates were very low during rainy days. The total areal evapotranspiration rate in the Bulang catchment is estimated to be 186.2 mm/year.

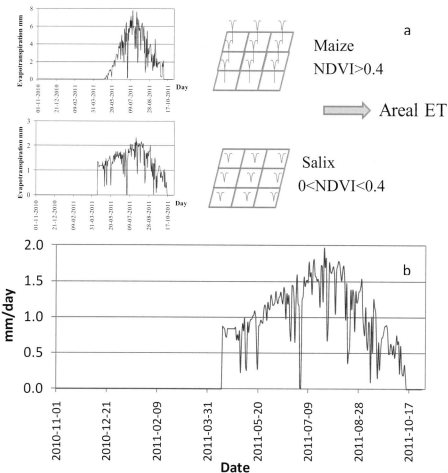

Figure 4.4 Up-scaling approach for areal evapotranspiration with sap flow measurements and NDVI values (a); and the estimated areal evapotranspiration (b).

The estimation of areal interception in the Bulang catchment followed the same approach as the estimation of the areal ET. The areal interception of 9.8 mm/year was obtained with the reference interception coefficient of 0.249 for salix (Yang, et al., 2008) and 0.373 for maize (LIN, et al., 2011).

Mu et al. (2011) created a data set of the evapotranspiration rates in 8 days intervals from MOD16 Global Terrestrial Evaporation Data Set. Figure 4.5 compares the areal ET rates of 8 days estimated from this study and Mu's data set. The larges differences are found during the winter period. ET values estimated from the remote sensing methods show significant different distribution compared with the ET values estimated in this study based on sap flow measurements in the experimental period, where ET rates are zero in the winter period from December to March. There should be no ET in winter since the daily average air temperature is below zero and there are no crops and the trees do not transpire in this period.

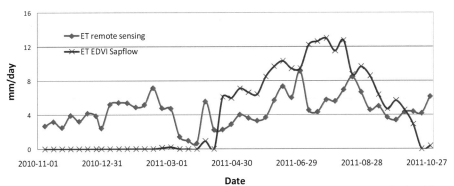

Figure 4.5 Comparison of evapotranspiration estimated by remote sensing and calculated by upscaling approach with NDVI and sap flow measurements from November 2010 to October 2011.

The change of the storage in the catchment is mainly based on the change of groundwater storage. Groundwater storage change can be estimated by the change of the groundwater levels multiplied by the porosity of the aquifer. Figure 4.6 shows that all groundwater levels at the end of the measuring period recovered back to the values at the beginning of the measuring period. Therefore, it can reasonably be assumed that there is no overall change of groundwater storage. The results of the water balance computation are shown in Figure 4.7 as daily water depth and are summarized in Table 4.3 as annual water depth in the catchment. The deep circulation in Table 3 refers to the regional groundwater flow to the Hailiutu River. The deep groundwater circulation was estimated in the water balance equation as the difference of total inflow minus total outflow. The net groundwater recharge equals to the precipitation minus total evapotranspiration.

Table 4.3 Annual water balance estimation in the Bulang catchment in mm/year.

Estimation method	Precipitation	ET	Net Recharge	Discharge	Deep Circulation	Change of Storage
Water Balance	214.8	196.0	18.8	12.6	6.2	0
Steady model	214.8	196.0	18.8	13.1	5.8	0

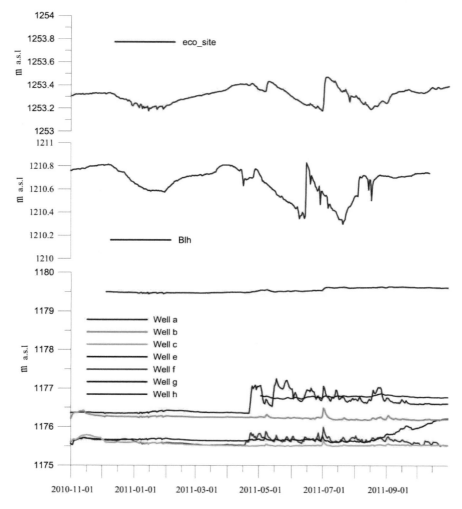

Figure 4.6 Observed groundwater levels in Bulang catchment.

4.3.1.2 Estimation of Groundwater Discharge

Groundwater discharge as the base flow in the Bulang catchment can be derived from the river discharge measurements during the experimental period. A base flow separation was carried out based on the analysis of the isotopic tracers in the river, groundwater, and the heavy rainfall occurred in July 2011 (Yang, et al., 2014). Results of the hydrograph separations illustrate that the pre-event component accounts for 74.8% of the total discharge during rainfall events. Compared to other graphical and mathematical hydrograph separation methods, the event based isotopic separation method provides more accurate estimation of the groundwater discharge. It is

75

estimated that 96.4% of total stream flow is composed of groundwater based on the event separation results from November 2010 to October 2011. Figure 4.8 illustrates the relations among the discharge, base flow, and the rainfall in the Bulang catchment.

4.3.1.3 Calibration of the steady groundwater model

Since the change of groundwater storage in the observed period from November 2010 to October 2011 can be neglected, a steady state groundwater flow model was constructed and calibrated first with annual average values of the net recharge, groundwater discharge, and groundwater levels. The purposes of the steady state model were to calibrate the hydraulic conductivity and to create initial conditions for the transient model. Average values of groundwater levels measured in nine observation wells (Figure 4.2) were used to compare model-calculated groundwater levels for the model calibration. The optimization code, PEST was used to optimize values of hydraulic conductivity in three zones so that the sum of squared differences between calculated and measured groundwater levels are minimized.

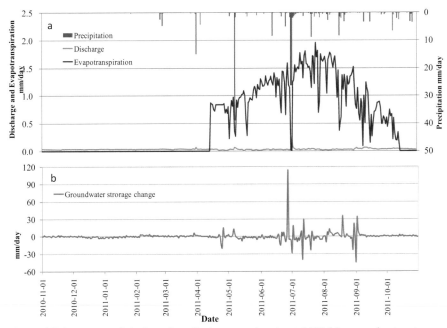

Figure 4.7 Average precipitation, river discharge, and estimated ET (a); groundwater storage change (b) in Bulang catchment from November 2010 to October 2011.

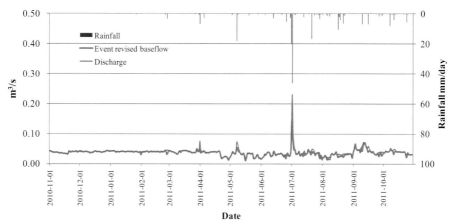

Figure 4.8 Relations among the observed discharge, separated baseflow, and the rainfall in the Bulang catchment.

4.3.1.4 Calibration of the transient groundwater model

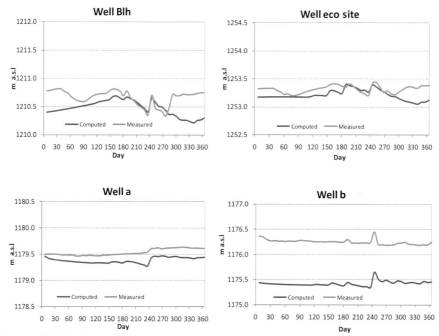

Figure 4.9 Fit of the computed groundwater levels to the measured ones in four wells in Bulang sub-catchment.

The purpose of the transient model is to simulate seasonal variations of groundwater levels and stream flow caused by varying precipitation and evapotranspiration values. The simulation period is one year from November 2010 to October 2011. A stress period of 8 days was chosen in line with estimated ET values from remote sensing data. The computed groundwater levels by the steady state model were used as initial groundwater heads in the transient model. The average values of daily precipitation and evapotranspiration values in every 8 days were used to compute transient net recharge values for the transient model. The specific yield values of three zones in the first model layer were optimized using PEST so that computed groundwater level series fit well with the measurement series. Figure 4.9 shows the match of the computed groundwater levels to the measurements in four observation wells. The RMSE values are 0.26 m, 0.14 m, 0.15 m, and 0.82 m for the computed groundwater levels at well Blh, eco-site, well a, and well b, respectively. The optimized values of specific yield are 0.29, 0.25 and 0.1, respectively, in the three parameter zones.

Figure 4.10 shows the comparison of the computed groundwater discharge to the stream and separated baseflow with the stream discharge measurements. The RMSE is 0.0075 m^3/s for the computed groundwater discharge. In general, the transient model computes more smooth discharges than the results of baseflow separation. Large differences in summer months may be caused by irrigation water use diverted from the river which is not considered in the model.

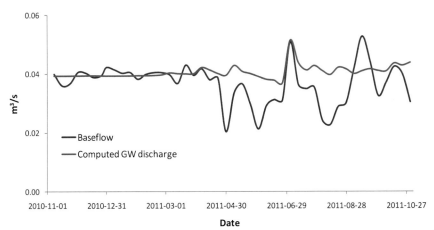

Figure 4.10 Comparison of simulated groundwater discharge to the Bulang River and the separated base flow from November 2010 to October2011.

4.3.1.5 Sensitivity of the transient groundwater model

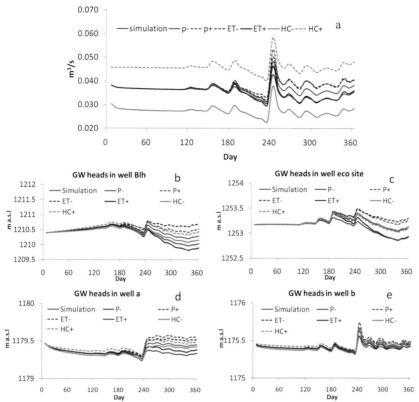

Figure 4.11 Sensitivity of computed stream discharge (a); and groundwater levels (b–e) to precipitation (P), evapotranspiration (ET), and hydraulic conductivity (HC); the sign (−) and (+) in the legend indicates the decreasing and increasing values by 26.5%.

Since the model was calibrated based on the one year measurements of groundwater levels and stream discharge and there are uncertainties in upscaling ET and estimating hydraulic conductivities, sensitivity analysis was performed to test sensitivities of computed groundwater levels and stream discharges to uncertainties in ET and hydraulic conductivity. In order to evaluate effects of inter-annual variations of precipitation, sensitivity of precipitation is also investigated. The coefficient of variation of the annual precipitation is 0.265 in the area. Therefore, precipitation, ET, and hydraulic conductivities were alternately increased and decreased by 26.5%, the transient model was run to compute the changes of groundwater levels and stream discharges while the other variables were kept fixed. The results are shown in Figure 4.11. The stream discharges are more sensitive to uncertainties in hydraulic conductivities while groundwater levels are more sensitive to changes in precipitation and ET values.

The calibrated transient groundwater flow model was used to simulate the scenarios. For each scenario, the model computes groundwater level changes, groundwater discharge to the Bulang River, deep groundwater circulation to the Hailiutu River, and change of groundwater storage.

4.3.1.6 Simulation of land use change scenarios

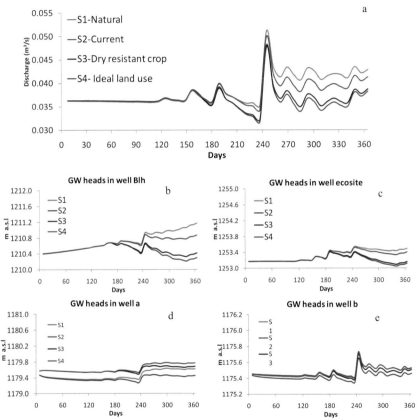

Figure 4.12 Simulated groundwater discharge to the Bulang River (**a**); and the calculated groundwater heads in well Blh (**b**); eco_site (**c**); well a (**d**);and well b (**e**) for four land use scenarios.

Annual water balance components for the four scenarios are summarized in Table 4.4. Time series of changes in discharge to the Bulang River and groundwater levels in four observation wells are plotted in Figure 4.12. The annual water balance computations show that net groundwater recharge increases significantly in scenario (1) and (4) when desert grasses were used for the protection of sand dunes. Discharge to the Bulang River increases slightly with increasing net recharge, while deep groundwater circulation discharging to the Hailiutu River remains constant. The benefit of increasing net recharge is largely in increasing groundwater storage, especially in scenarios (1) and (4). Increase of discharge to the Bulang River and increase of groundwater levels occur in the growing period from mid April to mid September (Figure 4.12). In the winter months, river discharge and groundwater levels are stable since there is neither precipitation nor evapotranspiration.

Table 4.4 Comparison of simulated water balance for four scenarios (mm/year).

Annual Components	Precipitation	ET	Net Recharge	Discharge	Deep Circulation	Change of Storage
Scenario 1	214.8	102.7	112.1	13.4	8.1	90.1
Scenario 2	214.8	196.0	18.8	12.7	8.0	−4.1
Scenario 3	214.8	189.2	25.6	12.8	8.0	3.9
Scenario 4	214.8	120.9	93.9	13.2	8.1	72.4

Annual water balance components for scenarios 2 and 4 under different hydrological years are summarized in Table 4.5. Time series of changes in discharge to the Bulang River and groundwater levels in four observation wells are plotted in Figure 4.13. The simulation results show that net groundwater recharge is increased significantly in normal and wet hydrological years as a result of increase of precipitation while evapotranspiration is only increased slightly comparing to the dry year. The increase of net recharge contributes largely to the increase of groundwater storage while discharge to the stream also increases. The increase of groundwater storage and stream discharges is more profound under the ideal land use scenario 4 compared to the current land use scenario 2. Under the ideal land use scenario, a healthy vegetation cover can be sustained in all years while water resources can be conserved for other social and economic uses.

Table 4.5 Comparison of simulated annual water balance components for scenarios 2 and 4 under different hydrological years (mm/year).

Annual Components	Precipitation	ET	Net Recharge	Discharge	Deep Circulation	Change of Storage
	Dry	214.8	196.0	18.8	12.7	8.0
Scenario 2	Normal	340.0	218.4	121.6	13.6	8.2
	Wet	420.0	280.5	139.5	13.8	8.4
	Dry	214.8	120.9	93.9	13.2	8.1
Scenario 4	Normal	340.0	136.7	203.3	14.1	8.3
	Wet	420.0	166.0	254.0	14.5	8.4

In order to evaluate the long-term impact of land use scenarios on groundwater discharges and groundwater levels in consideration of inter-annual variability of climate in the Bulang catchment, the calibrated transient model was extended to multiple years from 2000 to 2009. The month was chosen as stress period so that the transient model simulates 120 months. The monthly net groundwater recharge was estimated from monthly precipitations at Wushenqi station and estimated ET values. The monthly ET values for the current land use and ideal land use scenarios were estimated using the ratio of monthly ET to precipitation in the annual simulation of dry, normal and wet years. The simulation results are presented in Figure 4.14. In general, groundwater discharges and groundwater levels exhibit inter-annual variations. Groundwater levels and discharges were increased during the wet years of 2001 and 2002, and decreased during dry years of 2005 and 2006, and recovered during subsequent normal years. The depleted groundwater storage during the dry years can be restored during the normal and wet years. The alternating dry, normal and wet years will not cause degradation of vegetations since groundwater provides a reliable resource to sustain the vegetation. In comparison to the current land use, under the ideal land use scenario, groundwater levels in 4 observation wells are increased which lead to the increase of groundwater discharge to the river.

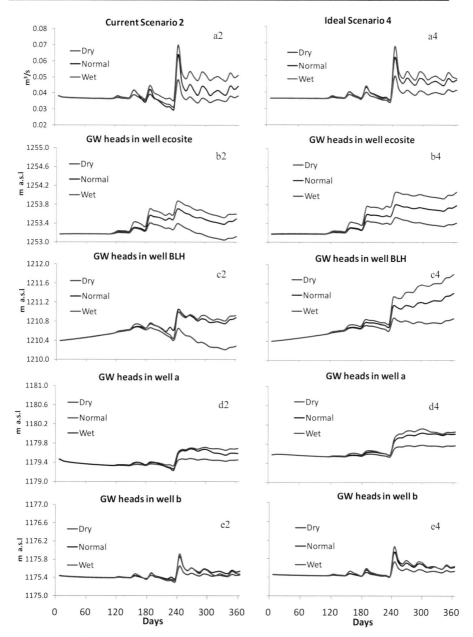

Figure 4.13 Simulated groundwater discharge to the Bulang River and calculated groundwater heads with current land use (a2–e2); and Ideal land use scenario (a4–e4) under dry, normal and wet hydrological years.

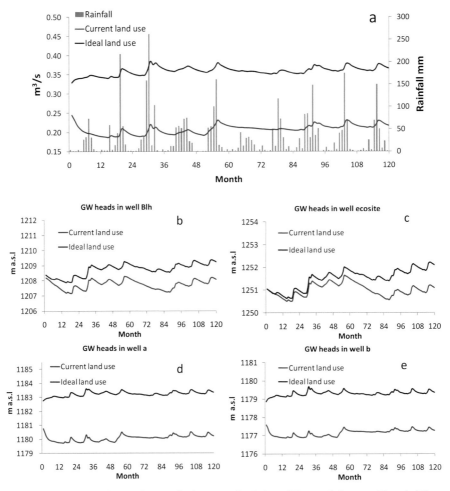

Figure 4.14 Simulated groundwater discharge to the Bulang River and the monthly rainfall at Wushenqi from 2000 to 2009 (a); the calculated groundwater heads in well Blh (b); eco_site (c); well a (d); and well b (e) for current and ideal land use scenarios from 2000 to 2009.

4.3.2 Discussion

Located in a semi-arid region, the Bulang catchment is covered by sand dunes. The hydrological processes are dominated by direct infiltration of precipitation and evapotranspiration. Direct surface runoff is limited and river discharge is maintained by groundwater discharge. Local authorities have implemented count mobilization measures against desertification by planting shrubs on unstable sand dunes, but large scale plantations of salix bushes to prevent desertification consumes too much water resources. The model simulation of current land use scenario in 2011 shows that in total 91% of the annual precipitation is consumed by the vegetation and crops in the catchment, the net groundwater recharge amounts to only 9% of the precipitation, which maintains stream discharge. However, for the ideal land use of planting desert grasses and dry resistant crops, only 56% of the annual precipitation would be consumed by dry resistant plants and crops and net groundwater recharge will be increased to 44% of the annual precipitation. In comparison with the result of scenario 3, the increase of net groundwater recharge is mainly contributed from planting desert grasses since the cropland area is limited to 3% of the land area in the Hailiutu River (Zhou, et al., 2015).

Inter-annual variations of precipitation have large impacts on the catchment water balance. The net groundwater recharge will be increased to 60.0% of the annual precipitation during normal and wet hydrological years for the ideal land use scenario. Groundwater storage and stream discharges are increased significantly during normal and wet years. The preliminary long-term simulations indicate that the depleted groundwater storage can be restored during normal and wet hydrological years while alternating dry, normal and wet hydrological years occur.

Although the study attempts estimating evapotranspiration rates using site measurements and upscaling with NDVI values, there are a number of limitations to be considered in the future research.

First, the site measurements of sap flow of salix bush, willow tree, and maize should be continued to obtain long-term reference evapotranspiration values for dry, normal and wet years. A field ecohydrological research site is under construction which includes large diameter lysimeters for measuring evapotranspiration of dominant plants and crops and net groundwater recharge at various groundwater table depths. Second, evapotranspiration of desert grasses (*Artemisia Ordosica*) and dry resistant crops (Sorghum and millet) should be measured. Eddy covariance technique (Zhao, et al., 2008) is most suited to measure evapotranspiration of grasses and crops at the plot scale. Third, the same NDVI map was used to compute the grid ET values for scenarios 2, 3, and 4 with the reference ET values of crops and plants. There are uncertainties in the estimated grid ET values with this upscaling method. Seasonal variations of NDVI values of crops and plants were not considered. ET values might change for different combinations of crops and plants in different scenarios. A more reliable relationship between seasonal NDVI and ET values could be established with long-term systematic measurements of plot scale ET values of dominant crops and plants with Eddy covariance technique.

The constructed groundwater model could be further improved by installing a number of monitoring wells in the recharge area in the northwestern part of the catchment. One more rain gauge at the northern boundary could provide better information about the spatial distribution of the rainfall input. Since most of the cropland is irrigated by means of groundwater abstraction, the investigation of the location and amount of groundwater abstraction should be implemented. Furthermore, the scattered small pools in local depressions with open water evaporation should be taken into account, where the evapotranspiration (ET) package of MODFLOW would be considered in the groundwater model.

The interaction between evapotranspiration and groundwater table depth was not simulated in the model. For scenarios with a significant increase of groundwater storage, groundwater table depth becomes shallower resulting higher evapotranspiration. This interaction can be partially simulated by using the ET package in MODFLOW.

The long-term simulations show that groundwater levels and discharges are relatively stable and fluctuate around long term means. A recent study by Yin et al. (Yin, 2015) simulated impacts of long-term climate variations on tree water use in the area. The result shows that trees did not suffer from water stress during the dry years because of the availability of groundwater for transpiration. The long-term monitoring of metrology, hydrology, and vegetation should be continued to ascertain the findings from model simulation.

4.4 Conclusion

Groundwater is the most important resource for local society and ecosystem protection in the semi-arid Bulang catchment. Groundwater maintains stream discharge and sustains vegetation growth. Net groundwater recharge, defined as precipitation minus evapotranspiration, is a good indicator of water resources availability in the area since dominant hydrological processes are direct infiltration of precipitation and evapotranspiration.

The simulation of the current land use in 2011 (being a dry year) indicates that nearly 91% of the annual precipitation is consumed by evapotranspiration of salix bushes and maize crops. Only 9% of the annual precipitation becomes net groundwater recharge which maintains a stable stream discharge. Although it is not possible to restore pristine land cover by desert grasses (scenario 1), an ideal land use (scenario 4) can be achieved by planting desert grasses for fixing sand dunes and dry resistant crops (sorghum and millet) for the local society. The simulation shows that the evapotranspiration consumes only 56% of the annual precipitation under the ideal land use scenario in dry year, and 40% in normal and wet years. The increase of net groundwater recharge is mainly contributed by planting desert grasses since the cropland area is limited.

The simulation of scenarios 2 and 4 under normal and wet hydrological years show significant increase of groundwater storage and slight increase of river discharges comparing to the dry year.

The long-term simulations show that groundwater levels and discharges are relatively stable and fluctuate around long-term means. The depleted groundwater storage during the dry years can be restored during normal and wet years.

The results of this study have relevant implications for water and ecosystem management in the catchment. In order to maintain river discharges and sustain a healthy growth of groundwater dependent vegetations, future land use changes should aim introducing dry resistant crops and desert grasses for vegetating sand dunes.

5. Groundwater and surface water interactions and impacts of human activities in the Hailiutu Catchment, Northwest China[*]

Abstract: The interactions between groundwater and surface water have been significantly affected by human activities in the semi-arid Hailiutu Catchment. A number of methods were used to investigate spatial and temporal interactions between groundwater and surface water. The isotopic and chemical analysis of surface water and groundwater samples indentified groundwater discharges to the river along the Hailiutu River. The mass balance equations were employed to estimate groundwater seepage rates along the river using the chemical profiles. The hydrograph separation method was used to estimate temporal variations of groundwater discharges to the river. A numerical groundwater model was constructed to simulate groundwater discharges along the river and analyze effects of water use in the catchment. The simulated seepage rates along the river compare reasonably well with the seepage estimates with the chemical profile in 2012. The impacts of human activities for river water diversion and groundwater abstraction on the river discharge were analyzed by calculating the differences between the simulated natural groundwater discharge and the measured river discharge. The water use in the Hailiutu River was increased from 1986 to 1991, reached its highest level from 1992 to 2000, and decreased from 2001 onwards. The reduction of river discharge might have negative impacts on the riparian ecosystem and the water availability for downstream users. The interactions between groundwater and surface water as well as the consequences of human impacts should be taken into account when implementing sustainable water resources management in the Hailiutu Catchment.

5.1 Introduction

In the last few decades more attention was given to the importance of interactions

---

[*]. This chapter is based on the paper Groundwater and surface water interactions and impacts of human activities in the Hailiutu Catchment, Northwest China. Yang, Z., Zhou, Y., Wenninger, J., Uhlenbrook, S., Wang, X., & Wan, L. (2017). Hydrogeological Journal. DOI 10.1007/s10040-017-1541-0.

between groundwater and surface water in water resources management and ecological studies (Brunke and Gonser, 1997, Winter, 1999, Sophocleous, 2002). The exchanges of surface water and water in the hyporheic zone as well as groundwater play very important roles for stream ecosystems (Findlay, 1995), hence in the consequence for ecology, river restoration and conservation (Boulton, et al., 2010). Understanding of the interactions between groundwater and surface water is also vital for the protection of groundwater-dependent ecosystems (Zhou, et al., 2013, Bertrand, et al., 2014). The surface water ecology can be affected by exchanges of groundwater through sustaining baseflow and moderating water-level fluctuations of groundwater-fed lakes (Hayashi and Rosenberry, 2002). The interaction between groundwater and surface water can be measured by a number of methods (Kalbus, et al., 2006, Brodie, et al., 2007), but their applicability depend on temporal and spatial scales.

Environmental tracers such as stable isotopes (deuterium and oxygen-18), radioactive isotopes (radon-222 or strontium), and chemical indices (ion concentrations and electrical conductivity) have been widely employed in determining the interactions between groundwater and surface water. Chemical and isotopic tracers have been used to characterize the interactions between groundwater and surface water in a karst area (Katz, et al., 1997) and in a lake catchment (Rautio and Korkka-Niemi, 2015), to estimate the groundwater recharge (Chen, et al., 2006, Gates, et al., 2008, Adomako, et al., 2010), to determine the surface water leakage into groundwater (Wang, et al., 2001), to identify the runoff processes (Wenninger, et al., 2008), to reflect the complex relations between the river and groundwater in an alluvial plain (Négrel, et al., 2003), and to identify locations and mechanisms of groundwater-surface water interaction in a pastorally dominated valley (Guggenmos, et al., 2011). The reliability of environmental tracer approaches to characterize the exchange between groundwater and surface water was concluded as efficient (Harvey, et al., 1996) and robust (Cook, 2013). However, as indicated, for instance, by Cook (2013), there are constrains related to these tracer methods such as (1) the assumption of steady-state conditions of river flows, (2) the needs for appropriate sampling resolution to reduce uncertainty in inflow rates, and (3) homogeneous solutes concentrations after mixing in wide and shallow rivers. McCallum et al. (2012) also emphasized that the concentration of the tracer in groundwater must be well defined, and the contrast between the concentration of the tracer in groundwater and the river water must be high for reducing the uncertainties when applying a tracer approach.

In order to reduce uncertainties, multiple methods with combinations of environmental tracers and other methods such as temperature measurements (Cook, et al., 2003), hydraulic techniques (Haria and Shand, 2004, Lamontagne, et al., 2005, Banks, et al., 2011), and different flow measurements (Cey, et al., 1998, McCallum, et al., 2012) have been adopted to investigate the interactions between groundwater and surface water recently. Banks et al. (2011) assessed the spatial and temporal aspects of the source and loss terms of the river and groundwater system using hydraulic, hydrochemical, and tracer-based techniques and determined how their relative magnitudes change along the river from the catchment headwaters towards the sea. McCallum et al. (2012) successfully determined both inflow and outflow components of groundwater with differential flow gauging and the sequential addition of electrical conductivity data as a natural tracer, chloride concentrations, and radon activity measurements. The piezometric surface monitoring method and environmental tracer methods were combined for identifying the temporal and spatial scales of groundwater–surface water exchange in the floodplain of the Murray River in Australia (Lamontagne, et al., 2005).

Spatial and temporal relations between groundwater and surface water might not only be controlled by natural processes, but also by anthropogenic factors. However, the quantification of human impacts on the interactions between groundwater and surface water is challenging because of their complexity. Hydro-geochemical methods have been used for the investigation of human impacts on groundwater. Lamontagne et al. (2005) pointed out that the flow regulation and water diversion for irrigation have considerable impacts on the exchange of surface water between the Murray River and its floodplains. The influence on the relationship between surface water and groundwater in a shallow Quaternary aquifer in the Mulde floodplain (Germany) was identified by geochemical methods (Petelet-Giraud, et al., 2007). In China, the groundwater quality evolution was used for investigating the groundwater recharge by leakage of the Tanghe wastewater reservoir (Wang, et al., 2014), identifying the impact of natural and anthropogenic factors on the groundwater chemistry in the Subei Lake basin (Liu, et al., 2015), and understanding the impact of enhanced anthropogenic pressure in the Dongguan area based on the comparison of hydrochemical data variations and land use changes during urbanization (Huang, et al., 2013). Durand et al. (2005) confirmed that human activities, in particular the pollution by the exploitation of the potash mines in the Alsace (France), mainly affected the Na and Cl concentrations of the river Rhine and the alluvial groundwater system based on the variations of major and trace elements.

Numerical modeling approaches have been carried out for studying the interactions between groundwater and surface water for transition zone water (Urbano, et al., 2006), importance of water balance in a mesoscale lowland river catchment (Krause and Bronstert, 2007, Krause, et al., 2007), in small catchments (Jones, et al., 2008, Guay, et al., 2013) with different scenarios (Gauthier, et al., 2009), water resources assessment (Henriksen, et al., 2008), evaluation of the impacts of best agricultural management practices (Cho, et al., 2010), and under the impacts of climate changes (Scibek, et al., 2007). Ala-aho et al. (2015) developed a fully integrated surface–subsurface flow model for the investigation of groundwater–lake interactions in an esker aquifer, which was verified with stable isotopes and airborne thermal imaging.

The objectives of this study are to quantify spatial and temporal interactions between groundwater and surface water, and to analyze impacts of human activities on these interactions in the Hailiutu River catchment located in the Erdos Plateau of China. Multiple methods were used to identify and quantify spatial and temporal interactions between groundwater and surface water. The isotopic and chemical analysis of surface water and groundwater samples indentified groundwater discharges to the river along the Hailiutu River. Mass balance equations were used to estimate groundwater seepage rates along the river using the chemical profiles. The hydrograph separation method was used to estimate temporal variations of groundwater discharges to the river. A numerical groundwater model was constructed to simulate groundwater discharges along the river and analyze effects of water use in the catchment. A better understanding of the interactions between groundwater and surface water as well as the consequences of human impacts provide science-based information for sustainable water resources management in the Hailiutu River catchment.

5.2 Materials and Methods

5.2.1 Study area

The semi-arid Hailiutu catchment lies in the middle part of the Yellow River basin in Northwest China with a total area of 2600 km² (Figure 5.1). The surface elevation of the Hailiutu catchment ranges from 1020 m in the Southeast to 1480 m above mean sea level in the Northwest. The perennial Hailiutu river flows from the Northwest to the Southeast and enters the Wuding River, which is the main tributary of the middle Yellow River (Yang, et al., 2012). The Hanjiamao hydrological station is located at the outlet of the catchment with a mean annual discharge of 2.62 m³/s from 1957 to 2012, while four meteorological stations inside and around the catchment were established in the 1950's providing rainfall and pan evaporation measurements. Most of the land surface is coved by sand dunes with scatted spots of desert bushes and grass except for the farmlands. The two reservoirs and several water diversions have been constructed along the main river channel and the Bulang tributary in order to provide irrigation water. The long-term annual average daily mean temperature is 8.1°C and the monthly mean daily air temperature is below zero in the winter time from November until March. The mean annual precipitation measured at the Wushenqi meteorological station was 340 mm/year. Precipitation mainly falls in June, July, August and September due to the monsoon climate. The mean annual pan evaporation (recorded with an evaporation pan with a diameter of 20 cm) is 2184 mm/year (Wushenqi metrological station, 1985–2004) (Yang, et al., 2014). The geological formations in the Bulang catchment mainly consist of four strata (1) the Holocene Maowusu sand dunes with a thickness from 0 to 30 m; (2) the upper Pleistocene Shalawusu formation (semi consolidated sandstone) with a thickness of 5 to ~90 m; (3) the Cretaceous Luohe sandstone with a thickness of 180 to ~330 m; and (4) the underlying impermeable Jurassic sedimentary formation (Yang, et al., 2015). A small drainage lake is formed by discharges from the Wushenqi country wastewater treatment plant, which is finally draining into the Hailiutu River.

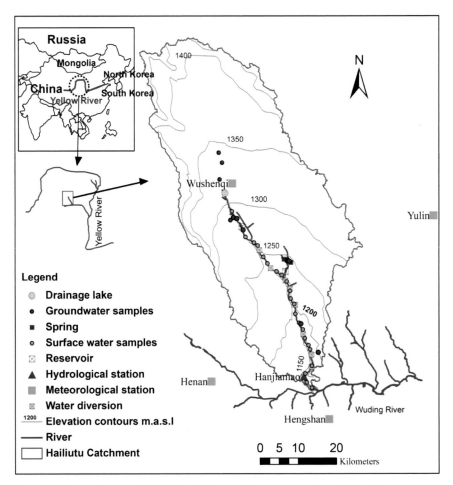

Figure 5.1 The catchment of the Hailiutu River and measurement/sample locations.

5.2.2 Discharge measurements

Daily river discharge at the Hanjiamao hydrological station had been observed by the Yellow River Conservation Commission (YRCC) at the outlet of the catchment since 1957. The updated ultrasonic river stage logger (type HW-1000c, Yellow River Hydroscience, Inc, Zhengzhou, China) combined with a concrete rectangle weir, were installed in 2001, which automatically records the water depth in front of the weir at the frequency of four times per day for normal and dry season and more frequently during flood events. The daily discharges have been converted from the water depth with verified rating curves. In order to validate the discharge measurements, an e+WATER L water level logger (type 11.41.53, Eijkelkamp Agrisearch Equipment, Giesbeek, The Netherlands) was installed at the weir, where water levels were

94

recorded with a frequency of 30 minutes. The recorded water depths from September 2010 to August 2011 are converted to discharges using a rating curve based on manual discharge measurements carried out with a current meter and the velocity-area method.

5.2.3 Chemical and isotopic sampling

Surface water samples of the drainage lake, river, and reservoir as well as groundwater samples of the deep and shallow aquifers, and the spring water samples in the Hailiutu catchment were taken for isotopic and chemical analysis during field reconnaissance studies in May 2010 and May 2012, respectively. Concentrations of major cations and anions of the 101 water samples such as Na^+, K^+, Ca^{2+}, Mg^{2+}, Cl^-, NO_3^{3-}, SO_4^{2-}, CO_3^{2-}, and SiO_2 were analyzed at the laboratory of the China University of Geosciences (Beijing) and the laboratory at the Xi'an Geological Survey Center with a Thermo IRIS Intrepid II (ICP-AES). Electrical conductivity (EC) values were measured with a portable water-quality multi-meter (18.28 and 18.21.SA temperature/conductivity meter, Eijkelkamp Agrisearch Equipment, Giesbeek, The Netherlands) in the field. Stable water isotopes oxygen-18 (^{18}O) and deuterium (^2H) of 216 samples were analyzed at UNESCO-IHE with a LRG DLT-100 isotope analyzer.

Values of the stable isotopes are reported as delta values in permil (‰) as a difference to the international reference Vienna Standard Mean Ocean Water (VSMOW), defined as:

$$\delta_{sample} = \left(\frac{R_{sample}}{R_{VSMOW}} - 1 \right) \times 1000 \qquad (5.1)$$

In which the δ_{sample} is the delta value [‰] for the isotopes in the sample; R_{sample} is the ratio of heavy to light isotopes in the sample, and R_{VSMOW} is the ratio of isotopes of the Vienna Standard Mean Ocean Water defined by the International Atomic Energy Agency (Klaus and McDonnell, 2013).

A linear relationship between deuterium and oxygen-18 was established for average global meteoric waters, which is well known as the Global Meteoric Water Line (GMWL) (Craig, 1961): $\delta D = 8\delta^{18}O + 10$. Monthly isotopes values in precipitation at the nearby Taiyuan, Baotou, and Yinchuan stations from the Global Network of Isotopes in Precipitation ((GNIP IAEA/WMO 2016)) are collected for providing localized relations between δD and $\delta^{18}O$ known as the Local Meteoric Water Line

(LMWL). The isotopic signatures of deuterium and oxygen-18 are often enriched in surface water bodies due to evaporative effects, while the isotopes remain unaffected in the groundwater system. The relation between the values of isotopes in different water bodies can be used to determining the interactions between groundwater and surface water. The isotopic and chemical profiles of the Hailiutu River were obtained from the analysis results of the river water samples with EC values, isotopic composition, and chemical components. The lengths from the origin of the Hailiutu River to the water samples sites range from 4 to 56.5 km (Figure 5.1).

5.2.4 Groundwater discharge estimation

Baseflow separation

    Previous studies (Yang, et al., 2012, Yang, et al., 2014) have shown that human impacts are the main driver for the stream flow changes and groundwater discharge to the river is the dominant hydrological process in the Hailiutu catchment. The significant changes in baseflow would indicate the human impacts on the interactions between groundwater and surface water. Thus, a quantitative analysis of the interactions between groundwater and surface water has been conducted with the hydrograph separation program HYSEP (Sloto and Crouse, 1996), which distinguish the groundwater discharge as baseflow from the total stream flow graphically based on the daily records. The baseflow index (BFI) is the long term ratio of baseflow to total stream flow, which was carried on the annual bases for groundwater discharge analysis.

Calculation with chemical profile

    The groundwater seepage along the Hailiutu River can be estimated with chemical profiles of the river and the discharge measurement at the Haijiamao hydrological station with mass balance equations. The mass balance equations of the chemical components and the total stream are expressed in Equation (5.2) and (5.3).

$$Q_{out}C_{out\_k} = Q_{in}C_{in\_k} + Q_gC_{g\_k} \qquad (5.2)$$

$$Q_{out} = Q_{in} + Q_g \qquad (5.3)$$

    Where the Q_{out}, Q_{in}, and Q_g is the stream flow at the downstream Hanjiamao station, the upstream inflow, and the total groundwater discharge in m^3/s; the C_{out_k}, C_{in_k}, and C_{g_k} are the concentrations of the kth chemical component in the river at the Hanjiamao station, the upstream inflow, and in the groundwater, respectively. The upstream inflow and total groundwater discharge can be calculated with the measurements at the Hanjiamao station, the analyzed chemical concentrations for river water samples, and the measured chemical concentrations of the groundwater.

The Hailiutu River can be divided into several segments according to the sample locations. The mass balance equations for the second segment from the upstream can therefore be written as:

$$Q_1 C_{1\_k} = Q_{in} C_{in\_k} + Q_{g1} C_{g\_k} \qquad (5.4)$$

$$Q_1 = Q_{in} + Q_{g1} \qquad (5.5)$$

Where the Q_1 and Q_{g1} are the discharge at the end of the segment and the groundwater seepage in the segment in m^3/s; C_{1_k} and C_{g_k} are the concentrations of the kth chemical component in the river at the end of the segment and groundwater seepage, respectively. The Q_1 and Q_{g1} can be solved by combining Equation (5.4) and (5.5):

$$Q_1 = Q_{in} \frac{C_{in\_k} - C_{g\_k}}{C_{1\_k} - C_{g\_k}} \qquad (5.6)$$

$$Q_{g1} = Q_1 - Q_{g1} \qquad (5.7)$$

The stream discharge and groundwater seepage at the remaining segments can be calculated consecutively with Equations (5.6) and (5.7).

5.2.5 Groundwater modeling

Model-set up

Based on the preliminary analysis of the interaction between groundwater and surface water in the Hailiutu catchment, it can be stated, that groundwater discharge to the river is the dominant hydrological process. A groundwater model with representation of physical and hydrological aspects for Hailiutu catchment was developed for simulating the interactions of the groundwater and surface water, impaction of the human activities, and the water use analysis. Two model layers were designed for simulating the groundwater aquifers according to the geological structure in the Hailiutu catchment. The upper Maowusu sand dunes together with Shalawush deposits were simulated with one unconfined aquifer layer and the underlying Cretaceous Luohe sandstone was simulated as the confined aquifer layer. All the lateral boundaries were taken as no-flow boundaries in line with the catchment water divide except for the southwest boundary at the bottom layer, where the outlet of the river valley was simulated as a head-dependent flow boundary for simulating deep groundwater flow entering the Wuding River. The Hailiutu River was simulated as a drain since groundwater always discharges to the stream. The numerical model code MODFLOW (McDonald and Harbaugh, 1988) was employed for the groundwater flow simulation. The uniform grid size of 500 by 500 m was used for the numerical model. This resulted in a numerical model grid consisting of 200 rows and 140 columns with 10489 active cells. The top elevation of the model is the surface elevation taken from the Digital Elevation Model with a resolution of 90 by 90 m. The bottom elevation of the model layers was interpolated from borehole data collected during the geological survey. The hydraulic conductivity for the upper layer was divided into five zones while a homogenous value was defined for the lower aquifer according to lithology and pumping test results. The horizontal hydraulic conductivities are 7.0, 5.0, 5.5, 2.0, and 1.2 m/d for the five zones of the top layer. The horizontal hydraulic conductivity for the sandstone layer was set to 0.5 m/d. The other hydrogeological parameters such as specific yield, effective porosity, and specific storage for both layers were determined based on the geological survey report (table 5.1). The General-Head Boundary (GHB) package of MODFLOW was used to simulate deep groundwater discharge to the Wuding River at the Southwestern boundary in the lower layer, where flow into or out of a GHB cell is calculated in proportion to the difference between the head in the model cell and the predefined stage of the Wuding River. The Drain package was used to simulate groundwater discharge to the stream since only groundwater discharges to the stream. The streambed elevations were used to specify the elevation of the drains. The Recharge and Evaporation package with precipitation and evaporation data measured at the meteorological stations were applied to simulate the recharge and evaporation processes in the groundwater model. The groundwater abstraction and the return flows of excess irrigation were not simulated since there are no data on these quantities. So, the model simulates natural groundwater flow without abstraction. The model boundaries are shown in Figure 5.2.

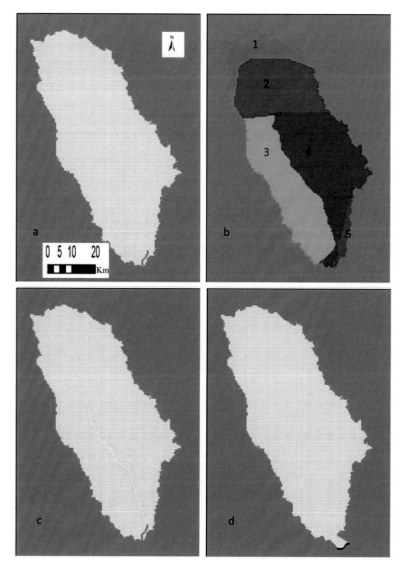

Figure 5.2 The model grid (a), five hydraulic conductivity zones at the top layer (b), Drain package at the top layer (c), Drain package and General-Head Boundary at the bottom layer (d).

Table 5.1. Hydraulic properties used in the models

Hydraulic parameters	Top layer	Bottom layer
Horizontal hydraulic conductivities (m/d)	Zone(1) 2.0 Zone(2) 5.0 Zone(3) 5.5 Zone(4) 7.0 Zone(5) 1.2	0.5
Specific yield (-)	0.05	0.01
Specific storage (m-1)	0.0001	0.0000299

Steady state model

According to a previous study about the flow regime shifts in the Hailiutu River (Yang, et al., 2012), the annual river flow from 1957 to 1967 was considered as quasi natural state with less influence from human activities such as the construction of the reservoirs and surface water diversion works along the river, and less abstraction of groundwater in the catchment for irrigation. The hydrological processes in that period represent the less interfered groundwater discharge to the Hailiutu River, which reflects the variations of the driving forces of precipitation and evaporation. Thus, a steady state flow model was calibrated first with the annual average values of rainfall, pan evaporation, stream discharges measured from 1957 to 1967 in order to determine the coefficients for inputting the recharge and evaporation by minimizing the deference between calculated and measured stream discharges. The simulation results and calibrated parameters can be used for the transient groundwater model.

Transient model

In order to estimate yearly groundwater discharges to the river, a transient groundwater flow model was constructed. The simulation period was from 1968 to 2012 with a stress period of a year. The computed groundwater heads from the steady state model were used as the initial heads. The values of hydraulic conductivity, specific yield and specific storage were estimated from the empirical values of the aquifer lithology. The transient stresses are groundwater recharge and evapotranspiration which were calibrated in the steady state model. In the simulation period, annual rainfall varies from 151 to 692 mm/a with a mean value of 340 mm/a, and open pan evaporation varies from 1753 to 2560 mm/a with a mean value of 2184 mm/a. The simulated groundwater discharge at each cell was reassembled into several river segments for the comparison with the groundwater seepage estimated with the chemical profile in 2012.

5.3 Results

5.3.1 Discharge measurements and baseflow separation

Comparison of measurements from YRCC station and Water L measurements

The comparison of measurements from the YRCC and the installed Water L with different frequencies during the experimental period at Hanjiamao is shown in figure 5.3 (a). The minimum, maximum, and mean discharges were measured by Water L are

0.74, 10.63, and 9.07 m^3/s, and those for YRCC are 2.17, 9.07, and 2.27 m^3/s, respectively. The peak flows recorded in September 2010 and July 2011 by the Water L with a high frequency were significantly larger than those obtained by YRCC with a lower frequency. The low flows recorded by the Water L during winter from November to March had comparably lower values than those measured by the YRCC. The baseflow separation was carried out with two daily discharge series data by a fix-interval, a sliding interval, and a local minimum method. Figure 5.3 (b) illustrates the average results of baseflow and discharge. The baseflow indices for the discharge measured by the YRCC and Water L equipment from September 2010 to August 2011 are 0.8563 and 0.8587, respectively.

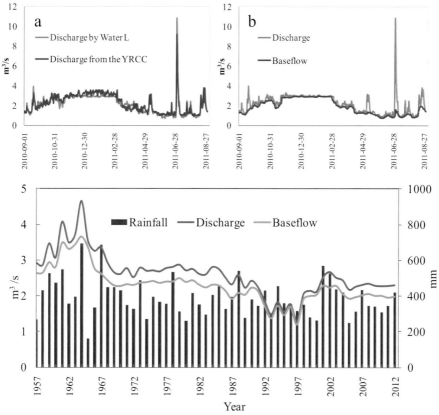

Figure 5.3 Comparison of discharge measurements by YRCC and Water L (a), results of baseflow separation and recorded discharge at Hanjiamao from September 2010 to August 2011 (b), and annual baseflow with stream discharge at Hanjiamao from 1957 to 2011.

Baseflow separation from 1957-2012

The baseflow separation had been conducted with the daily stream discharge records at the Hanjiamao hydrological station from 1957 to 2012. The annual baseflow and discharge for the Hailiutu river is shown in figure 5.3 (c). The mean annual baseflow index is 0.8553 with a minimum value of 0.7803 in 1967 and a maximum value of 0.9580 in 1996.

5.3.2 Isotopic and chemical analysis

Stable isotopic composition

Analysis results of the stable isotopes deuterium and oxygen-18 in surface water, reservoir water, spring water, groundwater water in the Hailiutu catchment, and the precipitation at nearby stations from GNIP are presented in figure 5.4 and table 5.2. The isotopic analysis includes samples from 31 surface water points, 2 reservoirs, 2 springs, 18 shallow groundwater boreholes, 1 deep groundwater borehole, and monthly average values of precipitation in winter and summer at the three nearby stations from GNIP, respectively. The local meteoric water line (LMWL; $\delta D = 7$ $\delta^{18}O + 0.94$) is determined based on the data collected from GNIP at the nearby three stations. The minimum, maximum, and mean values of stable isotopes of deuterium (in premil) in surface water/groundwater are -65/-69.66, -32.44/-52.98, and -57.09/-63.05, while the minimum, maximum, and mean values of oxygen-18 (in premil) in surface water/groundwater are -9.11/-9.97, -2.53/-7.69, and -7.57/-8.92, respectively. The minimum values of stable isotopes in the surface water and groundwater are almost the same except for those in the reservoir and springs. The maximum and the mean values of stable isotopes in surface water and groundwater are significantly different and indicate more enriched water samples in the surface water, following a clear evaporation line below the GMWL and LMWL.

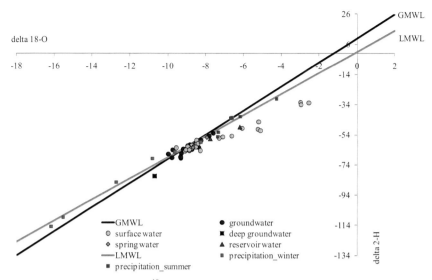

Figure 5.4 Values of δD and $\delta^{18}O$ in surface water, spring water, groundwater, and the precipitation in summer and winter at nearby stations in GNIP. The local meteoric water line (LMWL; $\delta D = 7 \delta^{18}O + 0.94$) is defined based on the data collected from GNIP.

Table 5.2 Summary of the stable isotopes for water samples in the Hailiutu catchment

Name	Number of samples	δD (permil)			$\delta^{18}O$ (permil)		
		minimum	maximum	mean	minimum	maximum	mean
Shallow groundwater	18	-69.66	-52.98	-63.05	-9.97	-7.59	-8.92
Deep groundwater	1		-81.42			-10.70	
Surface water	31	-65.00	-32.44	-57.09	-9.11	-2.53	-7.57
Reservoir	2	-56.75	-48.82	-52.78	-7.75	-6.18	-6.97
Spring water	2	-62.57	-55.76	-59.17	-8.67	-7.94	-8.31
Precipitation summer (GNIP) in three nearby stations	6	-16.16	-8.26	-12.18	-7.37	-4.23	-6.39
precipitation winter (GNIP) in three nearby stations	6	-114.83	-58.96	-83.83	-55.38	-30.19	-44.24

The chemical information for the groundwater and surface water samples are shown in figure 5.5. The chemistry of surface water samples are gradually changed from Na-SO$_4$ at the upstream drainage lake to Ca-HCO$_3$ type at more downstream locations. Groundwater samples show Ca-HCO$_3$ type in shallow groundwater wells and spring at the Bulang sub catchment. Samples of the 2 deep groundwater wells show a Ca-SO$_4$ type of water.

103

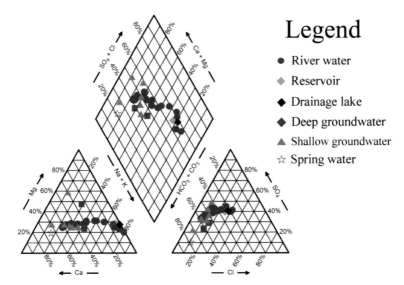

Figure 5.5 Piper diagram for river water, surface water, reservoir and drainage lake water, deep and shallow groundwater, and spring water samples taken in 2010 and 2012.

As shown in figure 5.6, groundwater samples from the upstream part of the catchment show higher SiO_2, NO_3^-, and Ca concentrations than those of the surface water. The higher Cl^-, Mg^{2+}, K^+, Na^+, and SO_4^{2-} in the surface water at the upstream part of the Hailiutu River were influenced by the discharge from the drainage lake with a high salinity level. The concentrations of these chemical compositions in the river water gradually decrease due to the discharge of groundwater with low concentrations along the course of the river. In the downstream section, chemical concentrations of river water are similar to those of the groundwater.

Profile of isotopes and chemistry

By using the analysis results of stable isotopes and chemical compositions of the water samples taken along the Hailiutu River in May 2012, the isotopic and chemical profiles of the Hailiutu River were plotted in figure 5.7. Stable isotopes of $\delta^{18}O$ and δD in the upstream part of the river starting from the drainage lake show high concentrations due to evaporation, and then the values gradually decrease to values similar to those of the groundwater samples. The chemical profiles of EC, Cl^-, SO_4^{2+}, Mg^{2+}, and Na show a similar pattern of changes: starting with high concentrations of the drainage lake and gradually decrease towards the groundwater concentrations due to groundwater discharge along the river. The profile of the Cl^- concentration was used to estimate groundwater discharge along the river since Cl⁻ is considered as a conservative tracer.

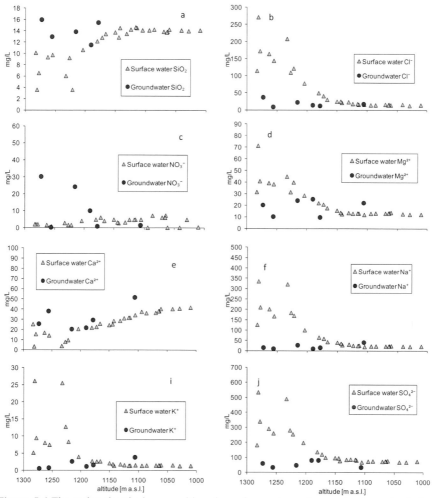

Figure 5.6 The major chemical compositions in surface water and groundwater water in the Hailiutu catchment in May 2012.

5.3.3 Estimation of groundwater discharge using the Cl⁻ chemical profile

The groundwater discharge as well as the groundwater seepage along the Hailiutu River was calculated with the Cl⁻ profile along the river and the measured discharge at the Haijiamao hydrological station in the experimental period using equations 5.2-5.7. Calculated minimum and maximum groundwater seepage rates was 0 m^3/s/meter at the 24-26, 32-34, 36-38 kilometer reaches and 0.0004658 m^3/s/meter at the 40-42 kilometer reach. The almost zero groundwater discharge rates were found at downstream 15 kilometers long reach from 40 kilometers on, which was calculated by comparable less fluctuation in the chemical profiles (figure 5.8).

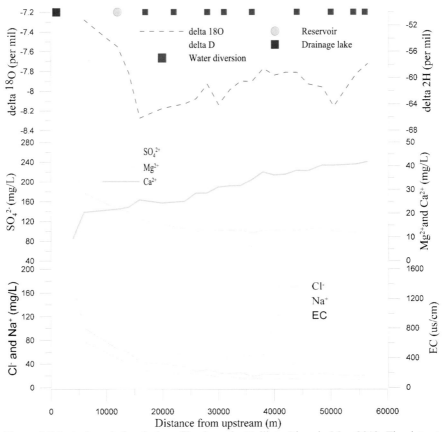

Figure 5.7 Isotopic and chemical profile along the Hailiutu River in May 2012. The dots at the top indicate the locations of drainage lake, Tuanjie reservoir, and water diversions at the Hailiutu river.

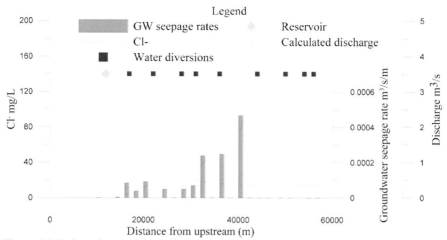

Figure 5.8 Estimated groundwater discharge and seepage rates along the Hailiutu River with the chemical profile and measured discharge at Hanjiamao hydrological station in May 2012.

5.3.4 Groundwater modelling results

Calibration of the groundwater model

The groundwater model was calibrated using a try and error method by changing the coefficients for inputting with the recharge and evaporation package in the steady state groundwater model. Because the observations of hydraulic heads were not available, the calibration of the steady-state model was achieved by means of comparing the computed and measured groundwater discharges. By minimizing the differences between simulated groundwater discharge and separated baseflow at Hanjiamao hydrological station from 1957 to 1967, the ratios for recharge of rainfall and maximum evaporation rates of measured pan evaporation were calibrated to be 0.5 and 0.24, respectively. The average annual baseflow separated by HYSEP and the simulated groundwater discharge at Hailiutu river are 3.06 and 3.063 m^3/s, respectively.

Simulation of natural groundwater discharge (baseflow)

The simulation of natural groundwater discharge from 1968 to 2012 to the Hailiutu River had been carried out with the transient groundwater model based on the calibrated parameters, simulated groundwater tables as the initial groundwater heads, and the measured model inputs of precipitation and pan evaporation data at meteorological stations. Figure 5.9 illustrates the measured river discharges, model-calculated groundwater discharges, and the separated baseflow from 1968 to

2012. The annual minimum, maximum, and mean values of measured river discharge, separated baseflow, and simulated groundwater discharge are 1.30/2.93/2.40, 1.20/2.49/2.10, and 2.58/2.99/2.81 m^3/s, respectively. The simulated groundwater discharges are obviously larger than the measured river discharges except for the period 1968 to 1985, which were caused by direct water diversion from the hydraulic works at the river and groundwater exploitation in the catchment to meet the demand of water consumption for local communities. Since the groundwater abstraction was not included in the transient model, the difference between the simulated natural groundwater discharges and the baseflow estimates provides gross estimates of water use in the Hailiutu catchment over the last decades. A part of the water use was the direct water diversion from the river, another part was groundwater abstraction. However, the abstracted groundwater would derive from a combination of (1) groundwater released from storage in the aquifer system and (2) capture of groundwater flow that would have discharged from the aquifer system through ET and contributions to stream baseflow. The effects of the abstraction can only be simulated in the transient model with the actual abstraction data.

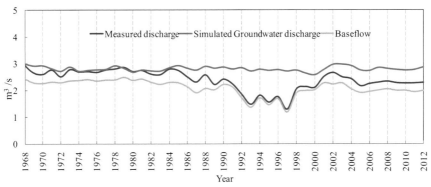

Figure 5.9 River discharge, simulated groundwater discharge, and the separated baseflow at Hanjiamao hydrological station from 1968 to 2012.

5.3.5 Comparison of seepage estimation and the simulated groundwater discharge

The comparison of calculated and simulated groundwater discharges as well as groundwater seepage rates along the Hailiutu River is shown in figure 5.10. The simulation results of annual discharges were taken in 2012 in accordance with the sampling period for chemical profile measured in May 2012. The discharge calculated by the chemical profile could be overestimated since peak flows occur in the summer from June to September. The total simulated and calculated groundwater discharges are 2.8503 and 2.85 m³/s at the outlet of the Hailiutu River. However, the calculated groundwater seepage rates are differing from the simulated seepage rates. The maximum rates for calculated and simulated groundwater seepage are 0.0004658 and 0.0001611 m³/s/m, respectively. The most significant difference between calculated and estimated groundwater seepage rates is located at the downstream part of the river.

5.4 Discussion

Daily discharges collected from the YRCC with a lower measuring frequency are less suited to detect high and low flows at the Hanjiamao gauging station during the experimental period from September 2010 to August 2011. The baseflow separation for both discharge series from YRCC and Water L result in similar baseflow indexes on an annual scale, which provides reliable groundwater discharges for the simulation of the groundwater model. As indicated in a previous study (Yang, et al., 2012), the stream flow regime as well as the separated baseflow have been shifted four times during the last decades by human activities on the local water resources management. Although the periods of flow regime shifts can be distinguished statistically, the quantification of the impact of human activities on the interactions between groundwater and surface water as well as the water resources management in the catchment remains complicated with respect to its temporal and spatial variability.

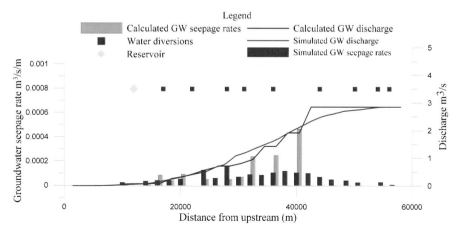

Figure 5.10 Comparison between the groundwater discharges calculated by the chemical profile and measured river discharges at Hanjiamao hydrological station and simulated by

groundwater flow model along the Hailiutu River.

The stable isotopes of deuterium and oxygen-18 of the groundwater are lying on the LMWL, which implies that the groundwater in the Hailiutu catchment have been direct recharged from precipitation. On the other hand, the isotopic compositions in the surface water samples are lying on an evaporation line, which indicates isotopic fractionation by evaporation effects at the reservoir and water diversions during the river course from upstream to downstream. The overlap of some surface water with groundwater and spring water samples in figure 5.4 represents the likely locations, where the groundwater and surface water are mixed or less affected by evaporation. Summer rainfall has more effect on groundwater/surface water since rainfall in summer accounts for more than 80% of the total precipitation. The regional groundwater flow in the deep aquifer travels longer before discharging into the river, thus, the higher concentration of cations and anions such as CO_3^-, Cl^-, and Mg^{2+} were found in the deep groundwater samples. Based on the comparison between results of the chemical analysis of deep groundwater, shallow groundwater, drainage lake water, and river water samples, the river water is mainly influenced by the shallow groundwater for most of the river water samples. This is shown in the piper diagram (figure 5.5), where the river water samples are located between the drainage lake sample and the shallow groundwater samples. There is a clear chemical evolution path from drainage lake ($Na-SO_4$ water) diluted with groundwater discharge to similar spring water ($Ca-HCO_3$ water) along the Hailiutu river. The drainage lake is the main source of Na^+, SO_4^{2-} and Cl^- released at the upstream part of the river, which also provide the opportunity for applying chemical mass balance method to estimate the groundwater discharge along the river with chemical profiles. The isotopic and chemical profiles along the Hailiutu river show that the interactions between groundwater and surface water in the Hailiutu catchment have been altered by construction of reservoirs and diversion works, irrigation on the farmlands, and other human activities releasing solutes through evaporation at reservoir/water diversions or solutes released from human activities.

The spatial interactions between groundwater and surface water as well as the human impacts can be investigated by evaluating the difference between the estimated groundwater discharge with chemical profiles and the simulated natural groundwater discharges. The advantages of estimating groundwater discharge using chemical profiles are (i) comparable less demanding with respect to hydro-geo-meteorological data, (ii) need shorter sampling periods, and (iii) can be effective in quantifying the human impacts. However, attention should be paid while interpreting the values for gaining river reaches as well as related to possible impacts of solutes originating from the top soil or human activities.

The water use in the last decades could be estimated by calculating difference between the simulated natural groundwater discharge, the measured river discharge, and the baseflow estimates on an annual basis. The large differences between the simulated natural groundwater discharge and the measured river discharge indicate the combined effects of river water diversion and groundwater abstraction. River water diversion has larger effect on peak flows in the raining season which coincides with the growing season of crops. Groundwater abstraction has double effects: it increases net groundwater recharge by reducing ET rates, and decreases groundwater discharge to the river by capturing natural groundwater discharge to the river. The water use in the Hailiutu River was increased from 1986 to 1991, reached its highest level from 1992 to 2000, and decreased from 2001 onwards. The reduction of river discharge would have negative impacts on the riparian ecosystem and the water use by downstream users. The interactions between groundwater and surface water as well as the consequences of human impacts should be taken into account when implementing sustainable water resources management in the Hailiutu Catchment.

5.5 Conclusion

Groundwater discharge accounts for 85.5% of the total amount of stream flow from 1957 to 2012 in the Hailiutu River. The interactions between groundwater and surface water in the Hailiutu catchment have been significantly affected by human activities such as the construction of reservoirs for water supply, direct water transfers from water diversions along the river, and groundwater abstraction for irrigation purposes. The groundwater and surface water in the catchment are originating from local precipitation as shown by the isotopic analysis results. Both stable isotopes of deuterium and oxygen-18 were enriched by evaporation effects at the drainage lake and at the hydraulic works. The chemical compositions of groundwater and surface water are also interfered by natural processes and human activities. There is a clear chemical evolution path from the drainage lake (Na-SO4 water) that is diluted with groundwater discharge (Ca-HCO3 water) along the Hailiutu River.

The spatial variation of human impacts on the interactions between groundwater and surface water was evaluated by comparing the estimated groundwater discharge rates with the chemical profile and the simulated groundwater discharges. The difference between groundwater discharge rates along the river estimated by the two methods shows that the groundwater discharge to the river can be affected by evaporation at the reservoirs, increase of the surface water stage at hydraulic works, and the solutes released by human activities. The most significant evidence of human impacts on the interactions between the groundwater and surface water is located at the reach with a distance of more than 42 km from the upstream, where almost no groundwater discharge was found by the artificial increased river stage using water diversion works for irrigation on the flood plain. The temporal variations of human impacts on the interactions between groundwater and surface water were investigated by comparing the simulated groundwater discharge and separated baseflow from 1957 to 2012 in the Hailiutu River. Furthermore, total amount of water consumption in the catchment can be estimated by the difference between the river flow, separated baseflow, and simulated groundwater discharge. The relations among the water consumption, the interactions between groundwater and surface water, and consequences of the human impacts were revealed. Intensified human impacts like an increased water use would lead to a decrease of river discharge and a higher percentage of groundwater in the Hailiutu River. Human impacts on the interactions between groundwater and surface water were rather constant from 1968 to 1985. The decline of river discharge and separated baseflow indicates that the human impacts increased from 1986 to 1991. During the period from 1992 to 2000, the river discharge, the separated baseflow, and the interactions between groundwater and surface water had been intensively influenced by human activities, for instance, through water transfers from the river for water supply and groundwater abstraction in the catchment. The human impacts have less effect on the interactions since 2001, the river discharge and estimated baseflow slightly recovered.

The understanding of human impacts on the interactions between groundwater and surface water can provide the valuable information for the ecosystem at flood plains and downstream areas, sustainable development in the catchment, but also for the water resources management in the Hailiutu catchment. A better understanding of the interactions between groundwater and surface water could be achieved through the application of a stream routing package in the groundwater model.

6 Conclusions and outlook

6.1 Conclusions

This study aimed to improve the understanding of interactions between groundwater and surface water in the semi-arid Hailiutu catchment in Erdos Plateau, Northwest China. A multi-disciplinary approach was adopted to quantify interactions between groundwater and surface water in the Hailiutu River catchment. The methods include a hydraulic approach, a hydrochemistry and isotope approach, a temperature approach, and a modelling approach. These methods were applied at various spatial and temporal scales from field surveys for chemical and isotopic profile along a tributary to the main river; historical and recent in-situ measurements, and numerical modelling at sub- and catchment scales. A number of conclusions can be drawn from this study.

(1) As a developing region in Northwest China, the temporal and spatial hydrological processes have been significantly altered in terms of river discharge, groundwater storage, and the interactions between groundwater and surface water over the past decades. The flow regime of the Hailiutu River has been changed dramatically from natural to overstressed status before 2000 and has slightly recovered since 2001. Four major shifts in the flow regime have been statistically detected in 1968, 1986, 1992 and 2001. The first period from 1957 to 1967 represents the natural variation of the river flow, with higher annual mean discharge, a large annual maximum discharge, a large standard deviation of mean monthly discharges, 6 months periodicity of two flow peaks (one in winter and one in summer) per year, and high daily flow variability. The flow regime in the second period from 1968 to 1985 and third period from 1968 to 1991 was gradually decreased by human activities, such as construction of reservoirs, water diversion works, and groundwater extraction for irrigation. After the most severe stress period of 1992 to 2000 with lowest flow rate, the river discharge and variations recovered in the fifth period of 2001 to 2007 due to the implementation of returning farmland to forest and grassland policy since 1999.

The cause-effect analysis identified that precipitation and air temperature had minor effects on flow regime changes, major flow regime shifts were caused by constructions of reservoirs and dikes for surface water diversion, groundwater extraction, and the land use changes.

(2) The interactions of groundwater and surface water was intensively investigated by means of multiple field measurements for an investigation period of one year in the Bulang sub-catchment, which indicates that groundwater recharge and discharges dominate the hydrological process. The systematic measurements of groundwater levels and temperature all indicate groundwater discharges to the river all year round. About 74.8% of the total river discharge was composed of groundwater discharge even during a heavy rainfall event. A novel estimation of groundwater seepage along the river reach was conducted with combined river discharge measurements and EC profile measurements under natural condition and constant injection. Results show high spatial variability of groundwater seepages along the river.

(3) Simulation of groundwater and river discharge variations under different land use scenarios in Bulang sub-catchment illustrates that evaporation from crops and bushes played very important roles in groundwater and surface water interactions under wet, normal, and dry conditions. Since nearly 91% of the annual precipitation is consumed by evaporation of *Salix* bushes and maize crops, and only 9% of it becomes net groundwater recharge, reduction of evaporation by changing land use is necessary to conserve water resources. Future land use changes should focus on planting desert bushes for stabilizing sand dunes and dry-resistant crops (sorghum and millet) for food production.

(4) The main characteristics of groundwater and surface water interactions in the Hailiutu River catchment is that groundwater constitutes major part of river discharges. Hydrograph separation with daily river discharges from 1957 to 2012 indicates that groundwater discharge accounts for nearly 86% of the total river discharges. Groundwater seepages along the river were calculated with mass balance equations of measured Cl concentrations and show large spatial variability. A transient

114

groundwater model could simulate historical natural groundwater discharges to the river and analyse human impacts on the river flow changes. The large differences between the simulated natural groundwater discharge and the measured river discharge indicate the combined effects of river water diversion and groundwater abstraction. River water diversion has larger effect on peak flows in the raining season which coincides with the growing season of crops. Groundwater abstraction has double effects: it increases net groundwater recharge by reducing ET rates, and decreases groundwater discharge to the river by capturing natural groundwater discharge to the river. The interactions between groundwater and surface water as well as the consequences of human impacts should be taken into account when implementing sustainable water resources management in the Hailiutu catchment.

(5) The findings from the thesis have important implications for water conservation and ecosystem management in the Hailiutu catchment. The water and ecosystem management priority should be to increase groundwater recharge by reducing evaporation loss. Plant species with less evaporation should be selected to vegetate sand dunes. Crops with lower irrigation demands should be promoted to reduce water abstraction. In the future, the Hailiutu River catchment must manage the groundwater recharge for water resource conservation and the maintenance of healthy vegetation.

6.2 Outlook and future research

There are a number of knowledge gaps in the present study that need to be further explored in the future research.

(1) While analysing the regime shifts in the Hailiutu River with the historical discharge observations, two flow peaks were found in the summer and winter before 1968. The summer flow peak was caused by the heavy rainfall in the summer due to the monsoon climate, which was vanished after construction of the reservoirs for river water diversion in the subsequent periods. However, the remaining winter flow peak was likely formed by groundwater discharge, which was also shifted to one month earlier. The possibly reason was due to the return flow of irrigation in the river valley, but irrigation started from April in this region for a long time. The generation

mechanism, behaviour, and implications of interactions between groundwater and surface water of winter flow peak should be investigated by intensive groundwater level monitoring in the basin during the winter.

(2) In this research, the interactions between groundwater and surface water at Bulang sub-catchment and the Hailiutu catchment have been studied at different temporal scale. However, the isotopic and hydrochemical aspects of interactions between groundwater and surface water on the flood plain in the river valley have not been taken into consideration, in which the complexity of interactions between groundwater and surface water has been increased by irrigation and drainage in farm fields. Further research could benefit from the natural and artificial tracer method during the irrigation period and beyond.

(3) Results of the hydrograph separations with isotopes illustrate that the pre-event water component accounts for 74.8% of the total discharge during heavy rainfall events in Bulang Tributary. The event-based hydrograph separation was conducted according to the ratios in the study. However, the percentage of pre-event water to the total discharge in response to the moderate rainfall events still needs to be determined and can be quantified by establishing routine and more frequent sampling during events for chemical and isotopic analysis at Hanjiamao hydrological station.

(4) The interesting hot spot that reflects huge human impacts on the interactions between groundwater and surface water is located at the south of the Wushenqi City, where an artificial wastewater discharge lake was formed by construction of a road. The northern part of the Hailiutu Basin should receive more attentions because it could be a local groundwater system that is hydraulically connected with surface water in a lake. A series of stand pipes in the stream bed can be installed to monitor the groundwater level and river stage with a data logger, which can directly distinguish the temporal variation into interactions between groundwater and surface water in the selected reach.

(5) A better understanding of the interactions between groundwater and surface water could be achieved through the application of a stream routing package in the groundwater model. The model parameters and boundary conditions could also be

optimized in the future research with spatially distributed groundwater level measurements. To improve the model simulation, the uncertainties of evaporation and groundwater recharge estimates can be reduced by establishing long-term observations of evaporation for different plants, groundwater levels (accompanied with appropriate registration methods), and river discharge in the catchment.

(6) Crucial data for groundwater model construction at the catchment scale are the water diversion for the irrigation in the river valley and groundwater abstraction for irrigation. Unfortunately, there were no historical data available in the present study. Therefore, data collection on the water use within the catchment should be conducted in the future through field observations and remote sensing interpretation. The amount of groundwater abstractions and river water diversions can significantly improve the model calibration and robustness.

References

Abu-Taleb AA, Alawneh AJ, Smadi MM (2007) Statistical analysis of recent changes in relative humidity in Jordan. Am J Environ Sci 3: 75-77

Adomako D, Maloszewski P, Stumpp C, Osae S, Akiti T (2010) Estimating groundwater recharge from water isotope ($\delta 2H$, $\delta 18O$) depth profiles in the Densu River basin, Ghana. Hydrological Sciences Journal–Journal des Sciences Hydrologiques 55: 1405-1416

Ahearn DS, Sheibley RW, Dahlgren RA, Anderson M, Johnson J, Tate KW (2005) Land use and land cover influence on water quality in the last free-flowing river draining the western Sierra Nevada, California. Journal of Hydrology 313: 234-247

Ala-aho P, Rossi PM, Isokangas E, Kløve B (2015) Fully integrated surface–subsurface flow modelling of groundwater–lake interaction in an esker aquifer: Model verification with stable isotopes and airborne thermal imaging. Journal of Hydrology 522: 391-406

Anderson JK, Wondzell SM, Gooseff MN, Haggerty R (2005) Patterns in stream longitudinal profiles and implications for hyporheic exchange flow at the HJ Andrews Experimental Forest, Oregon, USA. Hydrological Processes 19: 2931-2949

Anderson MP (2005) Heat as a ground water tracer. Ground water 43: 951-968

Arnell N, Reynard N (1996) The effects of climate change due to global warming on river flows in Great Britain. Journal of Hydrology 183: 397-424

Ayenew T, Kebede S, Alemyahu T (2008) Environmental isotopes and hydrochemical study applied to surface water and groundwater interaction in the Awash River basin. Hydrological Processes 22: 1548-1563

Banks E, Simmons C, Love A, Shand P (2011) Assessing spatial and temporal connectivity between surface water and groundwater in a regional catchment: Implications for regional scale water quantity and quality. Journal of Hydrology 404: 30-49

Batelaan O, De Smedt F, Triest L (2003) Regional groundwater discharge: phreatophyte mapping, groundwater modelling and impact analysis of land-use change. Journal of Hydrology 275: 86-108

Becker M, Georgian T, Ambrose H, Siniscalchi J, Fredrick K (2004) Estimating flow and flux of ground water discharge using water temperature and velocity. Journal of Hydrology 296: 221-233

Bellot J, Sanchez J, Chirino E, Hernandez N, Abdelli F, Martinez J (1999) Effect of different vegetation type cover on the soil water balance in semi-arid areas of south eastern Spain. Physics and Chemistry of the Earth, Part B: Hydrology, Oceans and Atmosphere 24: 353-357

Ben-Dor E, Goldshleger N, Braun O, Kindel B, Goetz A, Bonfil D, Margalit N, Binaymini Y, Karnieli A, Agassi M (2004) Monitoring infiltration rates in semiarid soils using airborne hyperspectral technology. International Journal of Remote Sensing 25: 2607-2624

Bertrand G, Siergieiev D, Ala-Aho P, Rossi P (2014) Environmental tracers and indicators bringing together groundwater, surface water and groundwater-dependent ecosystems: importance of scale in choosing relevant tools. Environmental earth sciences 72: 813-827

Bethenod O, Katerji N, Goujet R, Bertolini J, Rana G (2000) Determination and validation of corn crop transpirationby sap flow measurement under field conditions. Theoretical and Applied Climatology 67: 153-160

Bewket W, Sterk G (2005) Dynamics in land cover and its effect on stream flow in the Chemoga watershed, Blue Nile basin, Ethiopia. Hydrological Processes 19: 445-458

Bi YY, Zheng ZY (2000) The actual changes of cultivated area since the founding of new China. Resources Science 22: 8-12 (in Chinese with English abstract)

Bobba A, Bukata R, Jerome J (1992) Digitally processed satellite data as a tool in detecting potential groundwater flow systems. Journal of Hydrology 131: 25-62

Boulton AJ, Datry T, Kasahara T, Mutz M, Stanford JA (2010) Ecology and management of the hyporheic zone: stream-groundwater interactions of running waters and their floodplains. Journal of the North American Benthological Society 29: 26-40

Brodie R, Sundaram B, Tottenham R, Hostetler S, Ransley T (2007) An overview of tools for assessing groundwater-surface water connectivity. Canberra, Bureau of Rural Sciences, Australia.

Bronstert A, Niehoff D, Bürger G (2002) Effects of climate and land‐use change on storm runoff generation: present knowledge and modelling capabilities. Hydrological processes 16: 509-529

Brown AE, Zhang L, McMahon TA, Western AW, Vertessy RA (2005) A review of paired catchment studies for determining changes in water yield resulting from alterations in vegetation. Journal of Hydrology 310: 28-61

Brunke M, Gonser T (1997) The ecological significance of exchange processes between rivers and groundwater. Freshwater biology 37: 1-33

Buttle J (1994) Isotope hydrograph separations and rapid delivery of pre-event water from drainage basins. Progress in Physical Geography 18: 16-41

Cey EE, Rudolph DL, Parkin GW, Aravena R (1998) Quantifying groundwater discharge to a small perennial stream in southern Ontario, Canada. Journal of Hydrology 210: 21-37

Chen Y, Takeuchi K, Xu C, Xu Z (2006) Regional climate change and its effects on river runoff in the Tarim Basin, China. Hydrological Processes 20: 2207-2216

Chen Z, Nie Z, Zhang G, Wan L, Shen J (2006) Environmental isotopic study on the recharge and residence time of groundwater in the Heihe River Basin, northwestern China. Hydrogeology Journal 14: 1635-1651

Chiew FH, McMahon TA (2002) Modelling the impacts of climate change on Australian streamflow. Hydrological Processes 16: 1235-1245

Cho J, Mostaghimi S, Kang M (2010) Development and application of a modeling approach for surface water and groundwater interaction. Agricultural water management 97: 123-130

Choi W-J, Han G-H, Lee S-M, Lee G-T, Yoon K-S, Choi S-M, Ro H-M (2007) Impact of land-use types on nitrate concentration and δ 15 N in unconfined groundwater in rural areas of Korea. Agriculture, ecosystems & environment 120: 259-268

Christensen NS, Wood AW, Voisin N, Lettenmaier DP, Palmer RN (2004) The effects of climate change on the hydrology and water resources of the Colorado River basin. Climatic Change 62: 337-363

Cognard-Plancq A-L, Marc V, Didon-Lescot J-F, Normand M (2001) The role of forest cover on streamflow down sub-Mediterranean mountain watersheds: a modelling approach. Journal of Hydrology 254: 229-243

Cole ML, Kroeger KD, McClelland JW, Valiela I (2006) Effects of watershed land use on nitrogen concentrations and δ 15 nitrogen in groundwater. Biogeochemistry 77: 199-215

Conant B (2004) Delineating and quantifying ground water discharge zones using streambed temperatures. Groundwater 42: 243-257

Constantz J (1998) Interaction between stream temperature, streamflow, and groundwater exchanges in alpine streams. Water resources research 34: 1609-1615

Constantz J (2008) Heat as a tracer to determine streambed water exchanges. Water resources research 44 (4)

Constantz J, Stonestrom DA (2003) Heat as a tracer of water movement near streams. US Geological Survey circular: 1-96

Cook P, Favreau G, Dighton J, Tickell S (2003) Determining natural groundwater influx to a tropical river using radon, chlorofluorocarbons and ionic environmental tracers. Journal of Hydrology 277: 74-88

Cook PG (2013) Estimating groundwater discharge to rivers from river chemistry surveys. Hydrological Processes 27: 3694-3707

Costa MH, Botta A, Cardille JA (2003) Effects of large-scale changes in land cover on the discharge of the Tocantins River, Southeastern Amazonia. Journal of Hydrology 283: 206-217

Courault D, Seguin B, Olioso A (2005) Review on estimation of evapotranspiration from remote sensing data: From empirical to numerical modeling approaches. Irrigation and Drainage Systems 19: 223-249

Craig H (1961) Isotopic variations in meteoric waters. Science 133: 1702-1703

Dahmen E, Hall MJ (1990) Screening of hydrological data: tests for stationarity and relative consistency International Institute for Land Reclamation and Improvement

Deng X-P, Shan L, Zhang H, Turner NC (2006) Improving agricultural water use efficiency in arid and semiarid areas of China. Agricultural Water Management 80: 23-40

Didszun J, Uhlenbrook S (2008) Scaling of dominant runoff generation processes: Nested catchments approach using multiple tracers. Water resources research 44

Doherty J, Brebber L, Whyte P (1994) PEST: Model-independent parameter estimation. Watermark Computing, Corinda, Australia 122

Dong X, Zhang X, Yang B (1997) A preliminary study on the water balance for some sand land shrubs based on transpiration measurements in field condition. Acta Phytoecologica Sinica 21: 208-225

Dou L, Huang M, Hong Y (2009) Statistical Assessment of the Impact of Conservation Measures on Streamflow Responses in a Watershed of the Loess Plateau, China. Water resources management 23: 1935-1949 DOI 10.1007/s11269-008-9361-6

Drogue G, Pfister L, Leviandier T, El Idrissi A, Iffly J-F, Matgen P, Humbert J, Hoffmann L (2004) Simulating the spatio-temporal variability of streamflow response to climate change scenarios in a mesoscale basin. Journal of Hydrology 293: 255-269

Durand S, Chabaux F, Rihs S, Duringer P, Elsass P (2005) U isotope ratios as tracers of groundwater inputs into surface waters: example of the Upper Rhine hydrosystem. Chemical Geology 220: 1-19

Eckhardt K, Ulbrich U (2003) Potential impacts of climate change on groundwater recharge and streamflow in a central European low mountain range. Journal of Hydrology 284: 244-252

Farmer D, Sivapalan M, Jothityangkoon C (2003) Climate, soil, and vegetation controls upon the variability of water balance in temperate and semiarid landscapes: Downward approach to water balance analysis. Water Resources Research 39

Findlay S (1995) Importance of surface - subsurface exchange in stream ecosystems: The hyporheic zone. Limnology and oceanography 40: 159-164

Fohrer N, Haverkamp S, Eckhardt K, Frede H-G (2001) Hydrologic response to land use changes on the catchment scale. Physics and Chemistry of the Earth, Part B: Hydrology, Oceans and Atmosphere 26: 577-582

Ford CR, Hubbard RM, Kloeppel BD, Vose JM (2007) A comparison of sap flux-based evapotranspiration estimates with catchment-scale water balance. Agricultural and Forest Meteorology 145: 176-185

Fu G, Charles SP, Viney NR, Chen S, Wu JQ (2007) Impacts of climate variability on stream flow in the Yellow River. Hydrological Processes 21: 3431-3439

Furukawa Y, Inubushi K, Ali M, Itang A, Tsuruta H (2005) Effect of changing groundwater levels caused by land-use changes on greenhouse gas fluxes from tropical peat lands. Nutrient Cycling in Agroecosystems 71: 81-91

Gao H, Wang X, Yang G (2004) Sustainable Exploitation and Utilization of Water Resources in Erdos City. Site Investigation Science and Technology, 2, 005.

Gates JB, Edmunds W, Ma J, Scanlon BR (2008) Estimating groundwater recharge in a cold desert environment in northern China using chloride. Hydrogeology Journal 16: 893-910

Gauthier M, Camporese M, Rivard C, Paniconi C, Larocque M (2009) A modeling study of heterogeneity and surface water-groundwater interactions in the Thomas Brook catchment, Annapolis Valley (Nova Scotia, Canada). Hydrology and Earth System Sciences 13: 1583-1596

Ge Y, Boufadel MC (2006) Solute transport in multiple-reach experiments: Evaluation of parameters and reliability of prediction. Journal of Hydrology 323: 106-119

Gellens D, Roulin E (1998) Streamflow response of Belgian catchments to IPCC climate change scenarios. Journal of Hydrology 210: 242-258

Groeneveld DP (2008) Remotely-sensed groundwater evapotranspiration from alkali scrub affected by declining water table. Journal of Hydrology 358: 294-303

Groeneveld DP, Baugh WM, Sanderson JS, Cooper DJ (2007) Annual groundwater evapotranspiration mapped from single satellite scenes. Journal of Hydrology 344: 146-156

Guay C, Nastev M, Paniconi C, Sulis M (2013) Comparison of two modeling approaches for groundwater–surface water interactions. Hydrological Processes 27: 2258-2270

Guggenmos M, Daughney C, Jackson B, Morgenstern U (2011) Regional-scale identification of groundwater-surface water interaction using hydrochemistry and multivariate statistical methods, Wairarapa Valley, New Zealand. Hydrology and Earth System Sciences 15: 3383-3398

Guo H, Hu Q, Jiang T (2008) Annual and seasonal streamflow responses to climate and land-cover changes in the Poyang Lake basin, China. Journal of Hydrology 355: 106-122

Guidance of further institutional innovation on water management (2005) Ministry of Water Resources, China

Harbor JM (1994) A practical method for estimating the impact of land-use change on surface runoff, groundwater recharge and wetland hydrology. Journal of the American Planning Association 60: 95-108

Haria AH, Shand P (2004) Evidence for deep sub-surface flow routing in forested upland Wales: implications for contaminant transport and stream flow generation. Hydrology and Earth System Sciences 8: 334-344

Harvey JW, Wagner BJ, Bencala KE (1996) Evaluating the reliability of the stream tracer approach to characterize stream‐subsurface water exchange. Water Resources Research 32: 2441-2451

Hatch CE, Fisher AT, Ruehl CR, Stemler G (2010) Spatial and temporal variations in streambed hydraulic conductivity quantified with time-series thermal methods. Journal of Hydrology 389: 276-288

Hayashi M, Rosenberry DO (2002) Effects of ground water exchange on the hydrology and ecology of surface water. Groundwater 40: 309-316

He Y, Pu T, Li Z, Zhu G, Wang S, Zhang N, Xin H, Theakstone WH, Du J (2010) Climate change and its effect on annual runoff in Lijiang Basin-Mt. Yulong Region, China. Journal of Earth Science 21: 137-147

Hendricks Franssen H, Brunner P, Makobo P, Kinzelbach W (2008) Equally likely inverse solutions to a groundwater flow problem including pattern information from remote sensing images. Water Resources Research 44

Henriksen HJ, Troldborg L, Højberg AL, Refsgaard JC (2008) Assessment of exploitable groundwater resources of Denmark by use of ensemble resource indicators and a numerical groundwater–surface water model. Journal of Hydrology 348: 224-240

Hongve D (1987) A revised procedure for discharge measurement by means of the salt dilution method. Hydrological Processes 1: 267-270

Hou G, Liang Y, SU X, Zhao Z, Tao Z, Yin L, Yang Y, Wang X (2008) Groundwater systems and resources in the Ordos Basin, China. Acta Geologica Sinica (English Edition) 82: 1061-1069

Hou L, Wenninger J, Shen J, Zhou Y, Bao H, Liu H (2014) Assessing crop coefficients for Zea mays in the semi-arid Hailiutu River catchment, northwest China. Agricultural Water Management 140: 37-47

Hu LT, Chen CX, Jiao JJ, Wang ZJ (2007) Simulated groundwater interaction with rivers and springs in the Heihe river basin. Hydrological Processes 21: 2794-2806

Hu LT, Wang ZJ, Tian W, Zhao JS (2009) Coupled surface water–groundwater model and its application in the Arid Shiyang River basin, China. Hydrological Processes 23: 2033-2044

Hu Y, Maskey S, Uhlenbrook S (2012) Trends in temperature and rainfall extremes in the Yellow River source region, China. Climatic Change 110: 403-429

Hu Y, Maskey S, Uhlenbrook S, Zhao H (2011) Streamflow trends and climate linkages in the source region of the Yellow River, China. Hydrological Processes 25: 3399-3411

Huang G, Sun J, Zhang Y, Chen Z, Liu F (2013) Impact of anthropogenic and natural processes on the evolution of groundwater chemistry in a rapidly urbanized coastal area, South China. Science of the Total Environment 463: 209-221

Huang J, Zhou Y, Yin L, Wenninger J, Zhang J, Hou G, Zhang E, Uhlenbrook S (2014) Climatic controls on sap flow dynamics and used water sources of Salix psammophila in a semi-arid environment in northwest China. Environmental Earth Sciences: 1-13 DOI 10.1007/s12665-014-3505-1

IAEA/WMO (2016) Global Network of Isotopes in Precipitation. TheGNIP Database, Accessible at: http://www.iaea.org/water.Accessed 11 March 2016

Jeong CH (2001) Effect of land use and urbanization on hydrochemistry and contamination of groundwater from Taejon area, Korea. Journal of Hydrology 253: 194-210

Jha M, Pan Z, Takle ES, Gu R (2004) Impacts of climate change on streamflow in the Upper Mississippi River Basin: A regional climate model perspective. Journal of Geophysical Research: Atmospheres (1984–2012) 109

Jin X, Wan L, Zhang Y, Xue Z, Yin Y (2007) A study of the relationship between vegetation growth and groundwater in the Yinchuan Plain. Earth Science Frontiers 14: 197-203

Jin XM, Schaepman ME, Clevers JG, Su ZB, Hu G (2011) Groundwater depth and vegetation in the Ejina area, China. Arid Land Research and Management 25: 194-199

Jones J, Sudicky E, McLaren R (2008) Application of a fully‐integrated surface‐subsurface flow model at the watershed‐scale: A case study. Water Resources Research 44

Jothityangkoon C, Sivapalan M, Farmer D (2001) Process controls of water balance variability in a large semi-arid catchment: downward approach to hydrological model development. Journal of Hydrology 254: 174-198

Kalbus E, Reinstorf F, Schirmer M (2006) Measuring methods for groundwater? surface water interactions: a review. Hydrology and Earth System Sciences Discussions 10: 873-887

Kalma JD, McVicar TR, McCabe MF (2008) Estimating land surface evaporation: A review of methods using remotely sensed surface temperature data. Surveys in Geophysics 29: 421-469

Katz BG, Coplen TB, Bullen TD, Davis JH (1997) Use of chemical and isotopic tracers to characterize the interactions between ground water and surface water in mantled karst. Groundwater 35: 1014-1028

Kirchner JW (2003) A double paradox in catchment hydrology and geochemistry. Hydrological Processes 17: 871-874

Klaus J, McDonnell J (2013) Hydrograph separation using stable isotopes: Review and evaluation. Journal of Hydrology 505: 47-64

Krause S, Bronstert A (2007) The impact of groundwater–surface water interactions on the water balance of a mesoscale lowland river catchment in northeastern Germany. Hydrological Processes 21: 169-184

Krause S, Bronstert A, Zehe E (2007) Groundwater–surface water interactions in a North German lowland floodplain–implications for the river discharge dynamics and riparian water balance. Journal of Hydrology 347: 404-417

Krause S, Jacobs J, Bronstert A (2007) Modelling the impacts of land-use and drainage density on the water balance of a lowland–floodplain landscape in northeast Germany. Ecological Modelling 200: 475-492

Krein A, De Sutter R (2001) Use of artificial flood events to demonstrate the invalidity of simple mixing models/Utilisation de crues artificielles pour prouver l'invalidité des modèles de mélange simple. Hydrological sciences journal 46: 611-622

Lamontagne S, Leaney FW, Herczeg AL (2005) Groundwater – surface water interactions in a large semi‐arid floodplain: implications for salinity management. Hydrological Processes 19: 3063-3080

Landon MK, Rus DL, Harvey FE (2001) Comparison of instream methods for measuring hydraulic conductivity in sandy streambeds. Groundwater 39: 870-885

Langhoff JH, Rasmussen KR, Christensen S (2006) Quantification and regionalization of groundwater–surface water interaction along an alluvial stream. Journal of Hydrology 320: 342-358

Lavers D, Prudhomme C, Hannah DM (2010) Large-scale climate, precipitation and British river flows: Identifying hydroclimatological connections and dynamics. Journal of Hydrology 395: 242-255

Leblanc M, Favreau G, Tweed S, Leduc C, Razack M, Mofor L (2007) Remote sensing for groundwater modelling in large semiarid areas: Lake Chad Basin, Africa. Hydrogeology Journal 15: 97-100

Lewandowski J, Angermann L, Nützmann G, Fleckenstein JH (2011) A heat pulse technique for the determination of small‐scale flow directions and flow velocities in the streambed of sand‐bed streams. Hydrological Processes 25: 3244-3255

Li DH, Lv FY (2004) The function and economy of shrub in return farmland to forest and grass plan in Yulin. Shanxi forest: No 13 Green forum, (in Chinese)

Li H, Brunner P, Kinzelbach W, Li W, Dong X (2009) Calibration of a groundwater model using pattern information from remote sensing data. Journal of Hydrology 377: 120-130

Li LJ, Zhang L, Wang H, Wang J, Yang JW, Jiang DJ, Li JY, Qin DY (2007) Assessing the impact of climate variability and human activities on streamflow from the Wuding River basin in China. Hydrological Processes 21: 3485-3491

Li Y, Cui J, Zhang T, Okuro T, Drake S (2009) Effectiveness of sand-fixing measures on desert land restoration in Kerqin Sandy Land, northern China. Ecological engineering 35: 118-127

LIN D-j, ZHENG Z-c, ZHANG X-z, LI T-x, WANG Y-d (2011) Study on the Effect of Maize Plants on Rainfall Redistribution Processes. Scientia Agricultural Sinica 44: 2608-2615

Liu F, Song X, Yang L, Han D, Zhang Y, Ma Y, Bu H (2015) The role of anthropogenic and natural factors in shaping the geochemical evolution of groundwater in the Subei Lake basin, Ordos energy base, Northwestern China. Science of the Total Environment 538: 327-340

Lørup JK, Refsgaard JC, Mazvimavi D (1998) Assessing the effect of land use change on catchment runoff by combined use of statistical tests and hydrological modelling: case studies from Zimbabwe. Journal of Hydrology 205: 147-163

Love D, Uhlenbrook S, Twomlow S, Zaag Pvd (2010) Changing hydroclimatic and discharge patterns in the northern Limpopo Basin, Zimbabwe. Water SA 36: 335-350

Lowry CS, Walker JF, Hunt RJ, Anderson MP (2007) Identifying spatial variability of groundwater discharge in a wetland stream using a distributed temperature sensor. Water resources research 43

Lubczynski M W, Gurwin J (2005) Integration of various data sources for transient groundwater modeling with spatio-temporally variable fluxes—Sardon study case, Spain. Journal of Hydrology 306: 71-96

Lv J, Wang XS, Zhou Y, Qian K, Wan L, Eamus D, Tao Z (2013) Groundwater‐dependent distribution of vegetation in Hailiutu River catchment, a semi‐arid region in China. Ecohydrology 6: 142-149

Magilligan FJ, Nislow KH (2005) Changes in hydrologic regime by dams. Geomorphology 71: 61-78

Maheshwari B, Walker KF, McMahon T (1995) Effects of regulation on the flow regime of the River Murray, Australia. Regulated Rivers: Research & Management 10: 15-38

Marimuthu S, Reynolds D, La Salle C (2005) A field study of hydraulic, geochemical and stable isotope relationships in a coastal wetlands system. Journal of Hydrology 315: 93-116

Masih I, Uhlenbrook S, Maskey S, Smakhtin V (2011) Streamflow trends and climate linkages in the Zagros Mountains, Iran. Climatic Change: 1-22

Matheussen B, Kirschbaum RL, Goodman IA, O'Donnell GM, Lettenmaier DP (2000)

Effects of land cover change on streamflow in the interior Columbia River Basin (USA and Canada). Hydrological Processes 14: 867-885

McCallum JL, Cook PG, Berhane D, Rumpf C, McMahon GA (2012) Quantifying groundwater flows to streams using differential flow gaugings and water chemistry. Journal of Hydrology 416: 118-132

McDonald MG, Harbaugh AW (1984) A modular three-dimensional finite-difference ground-water flow model Scientific Publications Company Reston, VA, USA

McDonald MG, Harbaugh AW (1988) A modular three-dimensional finite-difference ground-water flow model

McDonnell J, Bonell M, Stewart M, Pearce A (1990) Deuterium variations in storm rainfall: Implications for stream hydrograph separation. Water resources research 26: 455-458

McLay C, Dragten R, Sparling G, Selvarajah N (2001) Predicting groundwater nitrate concentrations in a region of mixed agricultural land use: a comparison of three approaches. Environmental Pollution 115: 191-204

Mermoud A, Tamini T, Yacouba H (2005) Impacts of different irrigation schedules on the water balance components of an onion crop in a semi-arid zone. Agricultural Water Management 77: 282-295

Middelkoop H, Daamen K, Gellens D, Grabs W, Kwadijk JCJ, Lang H, Parmet BWAH, Schädler B, Schulla J, Wilke K (2001) Impact of Climate Change on Hydrological Regimes and Water Resources Management in the Rhine Basin. Climatic Change 49: 105-128 DOI 10.1023/a:1010784727448

Milliman J, Farnsworth K, Jones P, Xu K, Smith L (2008) Climatic and anthropogenic factors affecting river discharge to the global ocean, 1951-2000. Global and planetary change 62: 187-194

Mitchell D, Fullen M, Trueman I, Fearnehough W (1998) Sustainability of reclaimed desertified land in Ningxia, China. Journal of arid environments 39: 239-251

Molénat J, Gascuel‐Odoux C (2002) Modelling flow and nitrate transport in groundwater for the prediction of water travel times and of consequences of land use evolution on water quality. Hydrological processes 16: 479-492

Moore R (2004) Introduction to salt dilution gauging for streamflow measurement Part 2: Constant-rate injection. Streamline Watershed Management Bulletin 8: 11-15

Mu Q, Zhao M, Running SW (2011) Improvements to a MODIS global terrestrial evapotranspiration algorithm. Remote Sensing of Environment 115: 1781-1800

Münch Z, Conrad J (2007) Remote sensing and GIS based determination of groundwater dependent ecosystems in the Western Cape, South Africa. Hydrogeology Journal 15: 19-28

Neff R, Chang H, Knight CG, Najjar RG, Yarnal B, Walker HA (2000) Impact of climate variation and change on Mid-Atlantic Region hydrology and water resources. Climate Research 14: 207-218

Négrel P, Petelet-Giraud E, Barbier J, Gautier E (2003) Surface water–groundwater interactions in an alluvial plain: chemical and isotopic systematics. Journal of Hydrology 277: 248-267

Nie Z-l, Chen Z-y, Cheng X-x, Hao M-l, Zhang G-h (2005) The chemical information of the interaction of unconfined groundwater and surface water along the Heihe River, Northwestern China. Journal of Jilin University(Earth Science Edition) 35: 48-53

Niehoff D, Fritsch U, Bronstert A (2002) Land-use impacts on storm-runoff generation: scenarios of land-use change and simulation of hydrological response in a meso-scale catchment in SW-Germany. Journal of Hydrology 267: 80-93

Oxtobee J, Novakowski K (2002) A field investigation of groundwater/surface water interaction in a fractured bedrock environment. Journal of Hydrology 269: 169-193

Paulsen RJ, Smith CF, O'Rourke D, Wong TF (2001) Development and evaluation of an ultrasonic ground water seepage meter. Ground water 39: 904-911

Payn R, Gooseff M, McGlynn B, Bencala K, Wondzell S (2009) Channel water balance and exchange with subsurface flow along a mountain headwater stream in Montana, United States. Water resources research 45

Petelet-Giraud E, Negrel P, Gourcy L, Schmidt C, Schirmer M (2007) Geochemical and isotopic constraints on groundwater–surface water interactions in a highly anthropized site. The Wolfen/Bitterfeld megasite (Mulde subcatchment, Germany). Environmental pollution 148: 707-717

Pettitt A (1979) A non-parametric approach to the change-point problem. Applied statistics: 126-135

Poff NLR, Bledsoe BP, Cuhaciyan CO (2006) Hydrologic variation with land use across the contiguous United States: geomorphic and ecological consequences for stream ecosystems. Geomorphology 79: 264-285

Rautio A, Korkka-Niemi K (2011) Characterization of groundwater-lake water interactions at Pyhajarvi, a lake in SW Finland. Boreal environment research 16: 363-380

Rautio A, Korkka-Niemi K (2015) Chemical and isotopic tracers indicating groundwater/surface-water interaction within a boreal lake catchment in Finland. Hydrogeology Journal 23: 687-705

Richter BD, Baumgartner JV, Powell J, Braun DP (1996) A method for assessing hydrologic alteration within ecosystems. Conservation Biology 10: 1163-1174

Rientjes THM, Haile AT, Mannaerts CMM, Kebede E, Habib E (2010) Changes in land cover and stream flows in Gilgel Abbay catchment, Upper Blue Nile basin – Ethiopia. Hydrol Earth Syst Sci Discuss 7: 9567-9598 DOI 10.5194/hessd-7-9567-2010

Rodgers P, Soulsby C, Petry J, Malcolm I, Gibbins C, Dunn S (2004) Groundwater - surface - water interactions in a braided river: a tracer - based assessment. Hydrological Processes 18: 1315-1332

Rodionov S (2005) A brief overview of the regime shift detection methods Large-Scale Disturbances (Regime Shifts) and Recovery in Aquatic Ecosystems: Challenges for Management Toward Sustainability ed V Velikova and N Chipev UNESCO-ROSTE/BAS Workshop on Regime Shifts (Varna, Bulgaria, 14¨C16 June 2005.

Rodionov S, Overland JE (2005) Application of a sequential regime shift detection method to the Bering Sea ecosystem. ICES Journal of Marine Science: Journal du Conseil 62: 328-332

Rodionov SN (2004) A sequential algorithm for testing climate regime shifts. Geophys Res Lett 31: 9204

Rosenberry D, Pitlick J (2009) Local-scale spatial and temporal variability of seepage in a shallow gravel-bed river. Hydrological Processes 23: 3306-3318

Rosenberry DO, LaBaugh JW (2008) Field techniques for estimating water fluxes between surface water and ground water Geological Survey (US).

Rosenberry DO, Morin RH (2004) Use of an electromagnetic seepage meter to investigate temporal variability in lake seepage. Groundwater 42: 68-77

Ruehl C, Fisher A, Hatch C, Huertos ML, Stemler G, Shennan C (2006) Differential gauging and tracer tests resolve seepage fluxes in a strongly-losing stream. Journal of Hydrology 330: 235-248

Salama R, Hatton T, Dawes W (1999) Predicting land use impacts on regional scale groundwater recharge and discharge. Journal of Environmental Quality 28: 446-460

Scanlon BR, Reedy RC, Stonestrom DA, Prudic DE, Dennehy KF (2005) Impact of land use and land cover change on groundwater recharge and quality in the southwestern US. Global Change Biology 11: 1577-1593

Schmidt C, Bayer-Raich M, Schirmer M (2006) Characterization of spatial heterogeneity of groundwater-stream water interactions using multiple depth streambed temperature measurements at the reach scale. Hydrology and Earth System Sciences Discussions 10: 849-859

Schmidt C, Conant Jr B, Bayer-Raich M, Schirmer M (2007) Evaluation and field-scale application of an analytical method to quantify groundwater discharge using mapped streambed temperatures. Journal of Hydrology 347: 292-307

Scibek J, Allen DM, Cannon AJ, Whitfield PH (2007) Groundwater–surface water interaction under scenarios of climate change using a high-resolution transient groundwater model. Journal of Hydrology 333: 165-181

Scott RL, James Shuttleworth W, Goodrich DC, Maddock III T (2000) The water use of two dominant vegetation communities in a semiarid riparian ecosystem. Agricultural and Forest Meteorology 105: 241-256

Selker JS, Thévenaz L, Huwald H, Mallet A, Luxemburg W, Van de Giesen N, Stejskal M, Zeman J, Westhoff M, Parlange MB (2006) Distributed fiber - optic temperature sensing for hydrologic systems. Water resources research 42

Sener E, Davraz A, Ozcelik M (2005) An integration of GIS and remote sensing in groundwater investigations: a case study in Burdur, Turkey. Hydrogeology Journal 13: 826-834

Shaban A, Khawlie M, Abdallah C (2006) Use of remote sensing and GIS to determine recharge potential zones: the case of Occidental Lebanon. Hydrogeology Journal 14: 433-443

Silliman SE, Ramirez J, McCabe RL (1995) Quantifying downflow through creek sediments using temperature time series: One-dimensional solution incorporating measured surface temperature. Journal of Hydrology 167: 99-119

Sklash MG, Farvolden RN (1979) The role of groundwater in storm runoff. Journal of Hydrology 43: 45-65

Sloto RA, Crouse MY (1996) HYSEP, a computer program for streamflow hydrograph separation and analysis US Department of the Interior, US Geological Survey

Snyman H, Fouché H (1991) Production and water-use efficiency of semi-arid grasslands of South Africa as affected by veld condition and rainfall. Water SA 17: 263-268

Sophocleous M (2002) Interactions between groundwater and surface water: the state of the science. Hydrogeology Journal 10: 52-67

Steele-Dunne S, Lynch P, McGrath R, Semmler T, Wang S, Hanafin J, Nolan P (2008) The impacts of climate change on hydrology in Ireland. Journal of Hydrology 356: 28-45

Stisen S, Jensen KH, Sandholt I, Grimes DI (2008) A remote sensing driven distributed hydrological model of the Senegal River basin. Journal of Hydrology 354: 131-148

Taylor W (2000) Change-Point Analyzer 2.3 software package. Taylor Enterprises, Libertyville, Illinois.

Taylor WA (2000) Change-point analysis: a powerful new tool for detecting changes. preprint, available as http://www.variation.com/cpa/tech/changepoint html

Thodsen H (2007) The influence of climate change on stream flow in Danish rivers. Journal of Hydrology 333: 226-238

Timilsena J, Piechota T, Tootle G, Singh A (2009) Associations of interdecadal/interannual climate variability and long-term colorado river basin streamflow. Journal of Hydrology 365: 289-301

Tu M (2006) Assessment of the effects of climate variability and land use change on the hydrology of the Meuse river basin. UNESCO-IHE, Institute for Water Education

Tweed SO, Leblanc M, Webb JA, Lubczynski MW (2007) Remote sensing and GIS for mapping groundwater recharge and discharge areas in salinity prone catchments, southeastern Australia. Hydrogeology Journal 15: 75-96

Uhlenbrook S (2009) Climate and man-made changes and their impacts on catchments. Water Policy

Uhlenbrook S, Frey M, Leibundgut C, Maloszewski P (2002) Hydrograph separations in a mesoscale mountainous basin at event and seasonal timescales. Water resources research 38: 31-31-31-14

Uhlenbrook S, Hoeg S (2003) Quantifying uncertainties in tracer ‐ based hydrograph separations: a case study for two ‐ , three ‐ and five ‐ component hydrograph separations in a mountainous catchment. Hydrological Processes 17: 431-453

Urbano L, Waldron B, Larsen D, Shook H (2006) Groundwater–surfacewater interactions at the transition of an aquifer from unconfined to confined. Journal of Hydrology 321: 200-212

Vogt T, Schneider P, Hahn-Woernle L, Cirpka OA (2010) Estimation of seepage rates in a losing stream by means of fiber-optic high-resolution vertical temperature profiling. Journal of Hydrology 380: 154-164

Wagner BJ, Harvey JW (2001) Analysing the capabilities and limitations of tracer tests in stream-aquifer systemsImpact of Human Activity on Groundwater Dynamics: Proceedings of an International Symposium (Symposium S3) Held During the Sixth Scientific Assembly of the International Association of Hydrological Sciences (IAHS) at Maastricht, The Netherlands, from 18 to 27 July 2001 IAHS, pp. 191.

Wang L, Ni. G, Hu H (2006) Simulation of interactions between surface water and groundwater in Qin River basin. Journal of Tsinghua University (Science and Technology) 46: 1979-1986

Wang N, Zhang S, He J, Pu J, Wu X, Jiang X (2009) Tracing the major source area of the mountainous runoff generation of the Heihe River in northwest China using stable isotope technique. Chinese Science Bulletin 54: 2751-2757

Wang S, Tang C, Song X, Wang Q, Zhang Y, Yuan R (2014) The impacts of a linear wastewater reservoir on groundwater recharge and geochemical evolution in a semi-arid area of the Lake Baiyangdian watershed, North China Plain. Science of the Total Environment 482: 325-335

Wang SH, Xu X (2002) Analysis of non-staple economy in China. Ecological Economy: 17-20 (in Chinese)

Wang Y, (2008) The resources saving and environmental protection in construction of Erdos energy base, China sustainable development forum. (Vol. 1, pp. 38-42). (in Chinese)

Wang Y, Ma T, Luo Z (2001) Geostatistical and geochemical analysis of surface water leakage into groundwater on a regional scale: a case study in the Liulin karst system, northwestern China. Journal of Hydrology 246: 223-234

Waters P, Greenbaum D, Smart PL, Osmaston H (1990) Applications of remote sensing to groundwater hydrology. Remote Sensing Reviews 4: 223-264

Wels C, Cornett RJ, Lazerte BD (1991) Hydrograph separation: A comparison of geochemical and isotopic tracers. Journal of Hydrology 122: 253-274

Wenninger J, Uhlenbrook S, Lorentz S, Leibundgut C (2008) Identification of runoff generation processes using combined hydrometric, tracer and geophysical methods in a headwater catchment in South Africa/Identification des processus de formation du débit en combinat la méthodes hydrométrique, traceur et géophysiques dans un bassin versant sud-africain. Hydrological sciences journal 53: 65-80

Westhoff M, Savenije H, Luxemburg W, Stelling G, Van de Giesen N, Selker J, Pfister L, Uhlenbrook S (2007) A distributed stream temperature model using high resolution temperature observations

Winter TC (1995) Recent advances in understanding the interaction of groundwater and surface water. Reviews of Geophysics 33: 985-994

Winter TC (1999) Ground water and surface water: A single resource US Geological Survey

Woessner WW (2000) Stream and fluvial plain ground water interactions: rescaling hydrogeologic thought. Ground water 38: 423-429

Wolfe BB, Hall RI, Edwards TWD, Jarvis SR, Sinnatamby RN, Yi Y, Johnston JW (2008) Climate-driven shifts in quantity and seasonality of river discharge over the past 1000 years from the hydrographic apex of North America. Geophysical Research Letters 35: L24402

Xiao C (2006) Research on Transform Relationship Between Surface Water and Groundwater in Taoer River Fan. Journal of Jilin University(Earth Science Edition) 36: 234-239

Xu J (2011) Variation in annual runoff of the Wudinghe River as influenced by climate change and human activity. Quaternary International 244: 230-237

Xu YH, Zheng YF, Liu XL, Su FR (2009) Climate Change Analysis in Recent 50 Years in Ordos. Meteorology Journal of Inner Mongolia (in Chinese with English abstract)

Yan Y, Yang Z, Liu Q, Sun T (2010) Assessing effects of dam operation on flow regimes in the lower Yellow River. Procedia Environmental Sciences 2: 507-516

Yang T, Zhang Q, Chen YD, Tao X, Xu C, Chen X (2008) A spatial assessment of hydrologic alteration caused by dam construction in the middle and lower Yellow River, China. Hydrological Processes 22: 3829-3843

Yang X, Yan J, Liu B (2005) The analysis on the change characteristics and driving forces of Wuding River runoff. Advances in Earth Science 20: 637-642

Yang Z, Li X, Sun Y, Liu L, Zhang X, Ma Y (2008) Characteristics of rainfall interception and stemflow for Salix Psammophila in Maowusu sandland, Northwest China. Advances in Water Science 19: 693-698

Yang Z, Zhou Y, Wenninger J, Uhlenbrook S (2012) The causes of flow regime shifts in the semi-arid Hailiutu River, Northwest China. Hydrology and Earth System Sciences 16: 87-103

Yang Z, Zhou Y, Wenninger J, Uhlenbrook S (2012) The causes of flow regime shifts in the semi-arid Hailiutu River, Northwest China. Hydrology and Earth System Sciences 16: 87-103

Yang Z, Zhou Y, Wenninger J, Uhlenbrook S (2014) A multi-method approach to quantify groundwater/surface water-interactions in the semi-arid Hailiutu River basin, northwest China. Hydrogeology Journal 22: 527-541

Yang Z, Zhou Y, Wenninger J, Uhlenbrook S, Wan L (2015) Simulation of Groundwater-Surface Water Interactions under Different Land Use Scenarios in the Bulang Catchment, Northwest China. Water 7: 5959-5985

Yin L, Zhou. Y., Huang. J., Wenninger. J., Zhang. E., Hou. G., (2015) Interaction between groundwater and trees in an arid site: Potential impacts of climate variation and groundwater abstraction on trees. Journal of Hydrology 528: 435-448

Zampella RA, Procopio NA, Lathrop RG, Dow CL (2007) Relationship of Land‐Use/Land‐Cover Patterns and Surface‐Water Quality in The Mullica River Basin1. JAWRA Journal of the American Water Resources Association 43: 594-604

Zhang L, Dawes W, Walker G (2001) Response of mean annual evapotranspiration to vegetation changes at catchment scale. Water Resources Research 37: 701-708

Zhang Y-K, Schilling K (2006) Increasing streamflow and baseflow in Mississippi River since the 1940s: Effect of land use change. Journal of Hydrology 324: 412-422

Zhang ZZ, Mu XM, Wang F, Liu YL, Wang J (2009) Relationship Between Cultivated Land Change and Food Security in Yulin City. Research of Soil and Water Conservation (in Chinese with English abstract)

Zhao FF, Xu ZX, Zhang L, Zuo DP (2009) Streamflow response to climate variability and human activities in the upper catchment of the Yellow River Basin. Science in China Series E: Technological Sciences 52: 3249-3256

Zhao X, Huang Y, Jia Z, Liu H, Song T, Wang Y, Shi L, Song C, Wang Y (2008) Effects of the conversion of marshland to cropland on water and energy exchanges in northeastern China. Journal of Hydrology 355: 181-191

Zhou Y (1996) Sampling frequency for monitoring the actual state of groundwater systems. Journal of Hydrology 180: 301-318

Zhou Y, Wenninger J, Yang Z, Yin L, Huang J, Hou L, Wang X, Zhang D, Uhlenbrook S (2013) Groundwater–surface water interactions, vegetation dependencies and implications for water resources management in the semi-arid Hailiutu River catchment, China–a synthesis. Hydrology and Earth System Sciences 17: 2435-2447

Zhou Y, Wenninger J, Yang Z, Yin L, Huang J, Hou L, Wang X, Zhang D, Uhlenbrook S (2013) Groundwater–surface water interactions, vegetation dependencies and implications for water resources management in the semi-arid Hailiutu River catchment, China–a synthesis. Hydrology and Earth System Sciences, 17 (7): 2435–2447

Zhou Y, Yang Z, Zhang D, Jin X, Zhang J (2015) Inter-catchment comparison of flow regime between the Hailiutu and Huangfuchuan rivers in the semi-arid Erdos Plateau, Northwest China.

About the author

Zhi **YANG** was born in 1974 in Anhui Province, China. He graduated from the Hefei University of Technology, China with a BSc degree in Environmental Engineering in 1996. After that, he worked at the Institute of the Huai River Water Resources Protection as an environmental impact assessment consultant. He participated in the MSc programme Hydrology and Water Resources from 2002 to 2004 at UNESCO-IHE, Delft, The Netherlands. In December 2009, he entered the PhD programme at UNESCO-IHE (now called IHE Delft) in cooperation with China University of Geosciences (Beijing).

Selected publications:

Yang Z, Zhou Y, Wenninger J, Uhlenbrook S, Wang X, Wan L (2017) Groundwater and surface-water interactions and impacts of human activities in the Hailiutu catchment, northwest China. Hydrogeology Journal, 1-15. DOI :10.1007/s10040-017-1541-0

Yang Z, Zhou Y, Wenninger J, Uhlenbrook S, Wan L (2015) Simulation of groundwater-surface water interactions under different land use scenarios in the Bulang catchment, Northwest China. Water 7: 5959-5985. DOI: 10.3390/w9115959

Yang Z, Zhou Y, Wenninger J, Uhlenbrook S (2014) A multi-method approach to quantify groundwater/surface water-interactions in the semi-arid Hailiutu River basin, northwest China. Hydrogeology Journal 22(3): 527-541. DOI: 10.1007/s10040-013-1091-z

Zhou Y, **Yang Z**, Zhang D, Jin X, Zhang J (2015) Inter-catchment comparison of flow regime between the Hailiutu and Huangfuchuan rivers in the semi-arid Erdos Plateau, Northwest China. Hydrological Sciences Journal. 60: 688-705. DOI:10.1080/02626667.2014.892208

Zhou Y, Wenninger J, **Yang Z,** Yin L, Huang J, Hou L, Wang X, Zhang D, Uhlenbrook S

(2013)Groundwater–surface water interactions, vegetation dependencies and implications for water resources management in the semi-arid Hailiutu River catchment, China–a synthesis. Hydrology and Earth System Sciences, 17 (7):2435-2447. DOI: 10.5194/hess-17-2435-2013

Jin X, Guo R, Zhang Q, Zhou Y, Zhang D, Yang Z (2014) Response of vegetation pattern to different landform and water-table depth in Hailiutu River basin, Northwestern China. Environmental Earth Sciences. 71: 4889-4898. DOI: 10.1007/s 12665-013-2882-1

Lou Y, **Yang Z**, Zhang L, Li X (2013) Analysis of the temperature distribution in Bailianya reservoir, Proceedings of 2013 IAHR Congress, Tsinghua University Press, Beijing.

Yang Z, Liu H, Gu L (2013) Rational utilization on dredged sludge with heavy metals in Xinbian River conservancy project, Proceedings of 2013 IAHR Congress, Tsinghua University Press, Beijing.

Yang Z, Zhou Y, Wenninger J, Uhlenbrook S (2012) The causes of flow regime shifts in the semi-arid Hailiutu River, Northwest China. Hydrology and Earth System Sciences: 87-103. DOI:10.5194/hess-16-87-2012

Wang H, Zhang M, Zhu H, Dang X, **Yang Z,** Yin L (2012) Hydro-climatic trends in the last 50 years in the lower reach of the Shiyang River Basin, NW China. Catena 95: 33-41. DOI: 10.1016/j.catena.2012.03.00

Yang Z, Zhou Y, Wenninger J, Uhlenbrook S (2011) Analysis of stream flow characteristics of the Hailiutu River in the central Yellow River, China Water Resource and Environmental Protection (ISWREP), 2011 International Symposium on IEEE, pp. 783-786

Yin L, Hou G, Huang J, Li Y, Wang X, **Yang Z**, Zhou Y (2011) Using chloride mass-balance and stream hydrographs to estimate groundwater recharge in the Hailiutu River Basin, NW China Water Resource and Environmental Protection (ISWREP), 2011 International Symposium on IEEE, pp. 325-328

Netherlands Research School for the
Socio-Economic and Natural Sciences of the Environment

D I P L O M A

For specialised PhD training

The Netherlands Research School for the
Socio-Economic and Natural Sciences of the Environment
(SENSE) declares that

Zhi Yang

born on 9 August 1974 in Anhui, China

has successfully fulfilled all requirements of the
Educational Programme of SENSE.

Delft, 26 March 2018

the Chairman of the SENSE board

the SENSE Director of Education

Prof. dr. Huub Rijnaarts

Dr. Ad van Dommelen

The SENSE Research School has been accredited by the Royal Netherlands Academy of Arts and Sciences (KNAW)

K O N I N K L I J K E   N E D E R L A N D S E
A K A D E M I E   V A N   W E T E N S C H A P P E N

The SENSE Research School declares that Mr Zhi Yang has successfully fulfilled all
requirements of the Educational PhD Programme of SENSE with a
work load of 44.3 EC, including the following activities:

<u>SENSE PhD Courses</u>

o Research in context activity: 'Co-organizing workshop meeting on: 'the novelty of
   multidisciplinarity is the power of the creation: chemistry and isotopic methods on
   estimating groundwater seepage rates along the river'' (2011)
o Environmental research in context (2012)

<u>Other PhD and Advanced MSc Courses</u>

o Groundwater-river connectivity, UNESCO-IHE (2010)
o Groundwater dependent ecosystems, UNESCO-IHE (2010)
o Soil-vegetation-atmosphere transfer, UNESCO-IHE (2010)
o Ecohydrogeological modelling, UNESCO-IHE (2010)
o Management of water resources and ecosystems, UNESCO-IHE (2010)

<u>Management and Didactic Skills Training</u>

o Supervising two MSc students with theses entitled 'Study on characteristics and shift
   reasons of base flow for representative rivers on Ordos Plateau' (2011) and 'Analyzing
   impacts of land use change on hydrological regimes of Haihe river basin in China' (2012)

<u>Oral Presentations</u>

o *Analysis of stream flow characteristics of the Hailiutu River in the central Yellow River,
   China.* The 1st International Conference on Water Resource and Environment, 20-22
   May 2011, Xi'an, China
o *Use of hydrograph separation and groundwater flow model to assess water use in the
   Hailiutu River Catchment.* Seminar on Ecohydrology in Arid and Semiarid Regions,
   Partnership for Education and Research in Water and Ecosystem, Asian Facility for China,
   27-28 October 2011, Beijing, China
o *Impacts of human activities and climate change on flow regime shifts in the Hailiutu
   River Catchment.* Seminar on Ecohydrology in Arid and Semiarid Regions, Partnership for
   Education and Research in Water and Ecosystem, Asian Facility for China, 27-28 October
   2011, Beijing, China
o *Determination of groundwater-river connectivity and water balance with multiple field
   measurements in the Bulang sub-catchment of the Hailiutu River.* Seminar on
   groundwater-river interactions and vegetation dependency in water scarce
   environments , Partnership for Education and Research in Water and Ecosystem
   interactions, Asian Facility for China, 29 February 2012, Delft, The Netherlands

SENSE Coordinator PhD Education

Dr. Monique Gulickx

Printed and bound by CPI Group (UK) Ltd, Croydon, CR0 4YY

22/10/2024

01777647-0008